AutoCAD 3D Modeling

EXERCISE WORKBOOK

by
Steve Heather

Former Lecturer of
Mechanical Engineering &
Computer Aided Design.
AutoCAD Beta Tester

INDUSTRIAL PRESS, INC.

Industrial Press, Inc.

32 Haviland Street, Suite 3
South Norwalk, CT 06854
Phone: 203-956-5593
Toll-Free in USA: 888-528-7852
Fax: 203-354-9391
Email: info@industrialpress.com

Author: Steve Heather
Title: *AutoCAD© 3D Modeling Exercise Workbook, First Edition*
Library of Congress Control Number: 2017932682

© 2017 by Industrial Press, Inc.
All rights reserved. Published 2017.
Printed in the United States of America.

ISBN (print):	978-0-8311-3613-0
ISBN (ePDF):	978-0-8311-9428-4
ISBN (ePUB):	978-0-8311-9429-1
ISBN (eMobi):	978-0-8311-9430-7

Editorial Director: Judy Bass
Managing Editor: Laura Brengelman
Cover Designer: Janet Romano

industrialpress.com
ebooks.industrialpress.com

AutoCAD Books from Industrial Press

Beginning AutoCAD **2014**........................ ISBN 978-0-8311-3473-0
Beginning AutoCAD **2014** ebook............ ISBN 978-0-8311-9257-0
Advanced AutoCAD **2014**........................ ISBN 978-0-8311-3474-7
Advanced AutoCAD **2014** ebook............ ISBN 978-0-8311-9030-9

Beginning AutoCAD **2015**........................ ISBN 978-0-8311-3497-6
Beginning AutoCAD **2015** ebook............ ISBN 978-0-8311-9281-5
Advanced AutoCAD **2015**........................ ISBN 978-0-8311-3499-0
Advanced AutoCAD **2015** ebook............ ISBN 978-0-8311-9285-3

Beginning AutoCAD **2016**........................ ISBN 978-0-8311-3518-8
Beginning AutoCAD **2016** ebook............ ISBN 978-0-8311-9326-3
Advanced AutoCAD **2016**........................ ISBN 978-0-8311-3519-5
Advanced AutoCAD **2016** ebook............ ISBN 978-0-8311-9332-4

Beginning AutoCAD **2017**........................ ISBN 978-0-8311-3602-4
Beginning AutoCAD **2017** ebook............ ISBN 978-0-8311-9405-5
Advanced AutoCAD **2017**........................ ISBN 978-0-8311-3603-1
Advanced AutoCAD **2017** ebook............ ISBN 978-0-8311-9408-6

Beginning AutoCAD **2018**........................ ISBN 978-0-8311-3615-4
Beginning AutoCAD **2018** ebook............ ISBN 978-0-8311-9437-6
Advanced AutoCAD **2018**........................ ISBN 978-0-8311-3616-1
Advanced AutoCAD **2018** ebook............ ISBN 978-0-8311-9440-6

AutoCAD Pocket Reference
6th Edition, Releases 2013/2014 .... ISBN 978-0-8311-3484-6
AutoCAD Pocket Reference
6th Edition, ebook .... ISBN 978-0-8311-9170-2

AutoCAD Pocket Reference
7th Edition, Releases 2015/2016 .... ISBN 978-0-8311-3596-6
AutoCAD Pocket Reference
7th Edition, ebook .... ISBN 978-0-8311-9357-7

AutoCAD 3D Modeling .... ISBN 978-0-8311-3613-0
AutoCAD 3D Modeling ebook .... ISBN 978-0-8311-9428-4

For information about these and other bestselling titles, visit:
http://industrialpress.com and **http://ebooks.industrialpress.com**

Table of Contents

Introduction
About this Workbook..Intro-1
About the Author...Intro-1
Starting a New Drawing File Intro-2
Changing the Workspace Colors Intro-4
Changing the Drawing View Intro-7
Changing the Visual Style.......................................Intro-9
Status Bar Buttons...Intro-11
Using a Wheel Mouse...Intro-13
Application and Workspace Descriptions.................Intro-14

Lesson 1
Selecting a Basic 3D Tool.......................................1-2
Creating a Solid Box...1-6
Creating a Solid Cylinder 1-10
Creating a Solid Cone..1-13
Creating a Solid Sphere...1-15
Creating a Solid Pyramid 1-16
Creating a Solid Wedge...1-18
Creating a Solid Torus 1-22
Exercises 1-25

Lesson 2
Opening the Properties Palette................................2-2
Modify Solids Using the Properties Palette...............2-4
Modify Solids Using the Grips..................................2-7
Exercises 2-10

Lesson 3
Setting the Solid History System Variable 3-2
Chamfer the Edge of a 3D Solid Model 3-3
Modify an Existing Chamfer.....................................3-8
Remove an Existing Chamfer 3-10
Fillet the Edge of a 3D Solid Model..........................3-11
Modify an Existing Fillet..3-15
Remove an Existing Fillet 3-17
Exercises 3-18

Lesson 4
User Coordinate System (UCS)...............................4-2
Moving the UCS to a Temporary Location.................4-4
Rotating the UCS..4-8
Using the Dynamic UCS 4-10
Exercises 4-12

Lesson 5

Orbit Tool...5-2
Move Tool...5-4
Union Tool..5-6
Subtract Tool..5-12
Intersect Tool...5-16
Exercises..5-20

Lesson 6

3D Rotate Tool 6-2
3D Mirror Tool..6-6
3D Align Tool 6-8
Exercises..6-10

Lesson 7

Extrude Tool 7-2
Revolve Tool...7-9
Loft Tool..7-13
Sweep Tool..7-17
Exercises..7-22

Lesson 8

Shell Tool..8-2
Helix Tool..8-6
Exercises..8-10

Projects

Project 1
Stackable Junk Tray

Project 2
Ornate Balcony

Project 3
Working Platform

Project 4
Belt Roller Assembly

Appendixes
A 3D Printing
B Add a Printer / Plotter

Index

Final Notes about AutoCAD®

AutoCAD® 3D Modeling

EXERCISE WORKBOOK

Introduction

About this Workbook

This Workbook is designed for classroom instruction or self-study **and is suitable for both inch and metric users**. There are 8 lessons and 4 modeling projects.

Each lesson starts with step-by-step instructions on how to create 3-dimensional (3D) solid models followed by exercises designed for practicing the commands you learned within that lesson. The modeling projects are designed so that you can create complex 3D models by combining many of the commands you learned within those lessons.

Downloadable sample files are provided to accompany some of the lessons and modeling projects.

How to get the supplied sample files?

The file "**3d-modeling.zip**" should be downloaded from our website:

http://new.industrialpress.com/ext/downloads/acad/3d-modeling.zip

Enter the address into your web browser and the download will start automatically. Once the file has been downloaded you can unzip it to extract the sample files.

AutoCAD vs. AutoCAD LT

The LT version of AutoCAD has approximately 80 percent of the capabilities of the full version. It was originally created to be installed on the small hard drives that Laptops used to have. Hence, the name LT. (LT does not mean "Light"). In order to reduce the size of the program AutoCAD removed some of the high-end capabilities, such as Solid Modeling. As a result, AutoCAD LT is not suitable for this Workbook.

About the Author

Steve Heather is a former Lecturer of Mechanical Engineering and Computer Aided Design in England, UK. For the past 8 years he has been a Beta Tester for Autodesk®, testing the latest AutoCAD® software. Previous to teaching and for more than 30 years, he worked as a Precision Engineer in the Aerospace and Defense industries.

Steve can be contacted for questions or comments at: **steve.heather@live.com**

Starting a New Drawing File

This Workbook was created using AutoCAD® 2017. You may have an earlier version. There are some cosmetic changes to the main interface but the commands are mostly the same. During some of the lessons you will be asked to start a new drawing file using either **"acad.dwt"** for **inch** users, or **"acadiso.dwt"** for **metric** users. These are the standard template files that are automatically installed on your system when you install the AutoCAD software.

How to start a new drawing file.

1. Start AutoCAD.

2. Left click on the large "**A**" in the top left-hand corner.

3. Left click on **New**.

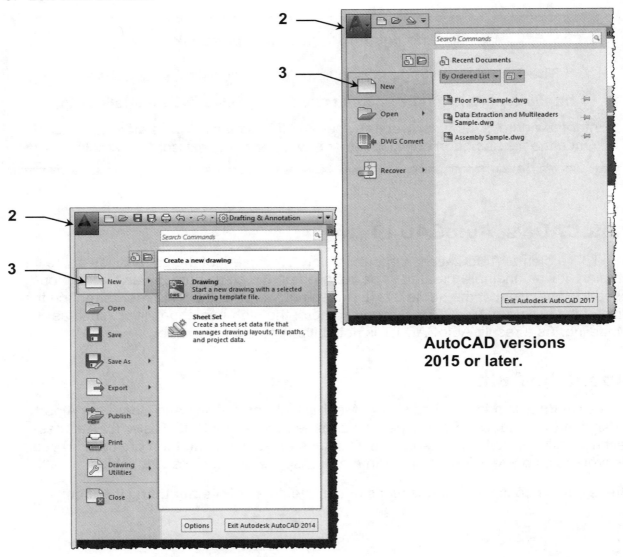

AutoCAD versions 2015 or later.

AutoCAD versions 2014 or earlier.

Continued on the next page...

Starting a New Drawing File....continued

4. In the **Select Template** dialog box, either click on **"acad.dwt"** if you are an **inch** user, or click on **"acadiso.dwt"** if you are a **metric** user.

5. Select **Open** if the drawing file hasn't already opened.

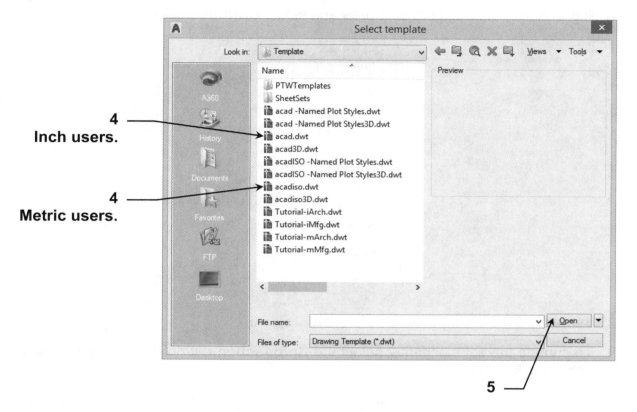

The drawing file will open showing you the **Workspace**. I have changed certain colors on the Workspace. If you wish to do the same refer to the next page for instructions.

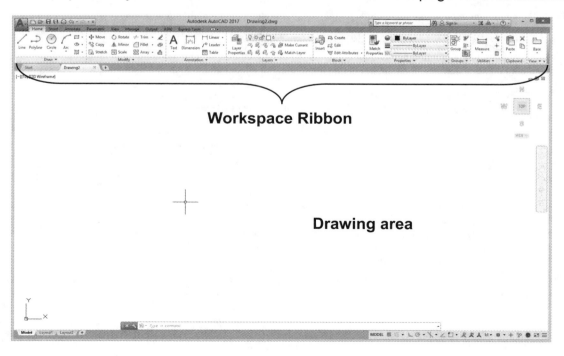

Changing the Workspace Colors

The default color scheme for AutoCAD is dark and the main drawing area is black. I prefer to have a light color scheme and a white drawing area. If you wish to change the color scheme or the drawing area color do the following.

1. Right click in the main drawing area and select **Options** from the list.

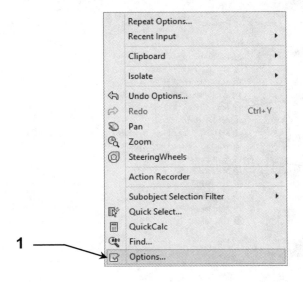

2. In the **Options** dialog box select the **Display** Tab.

3. To change the color scheme, select the **Color scheme** drop-down list and select either **Dark** or **Light**.

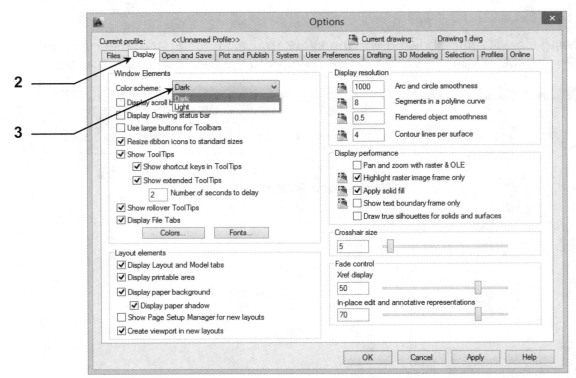

Continued on the next page...

Changing the Workspace Colors....continued

4. To change the color of the main drawing area select **Colors**.

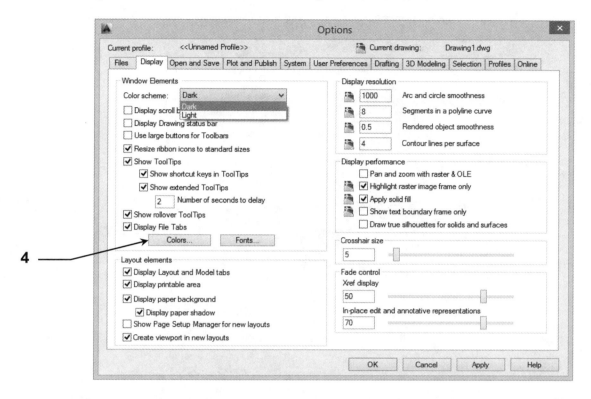

5. In the **Drawing Window Colors** dialog box select **2D Model space** (2D means 2-dimensional) in the **Context** window.

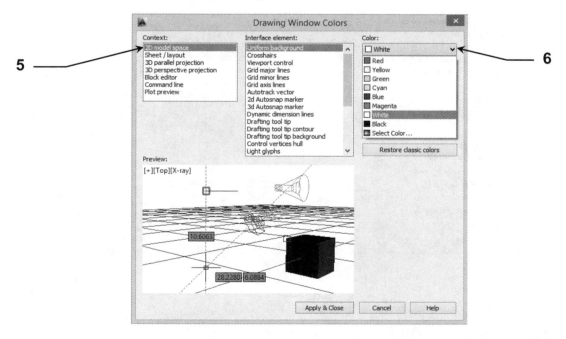

6. Select your preferred color in the **Color** drop-down list.

Continued on the next page...

Changing the Workspace Colors....continued

7. When you are happy with your choices select **Apply & Close**.

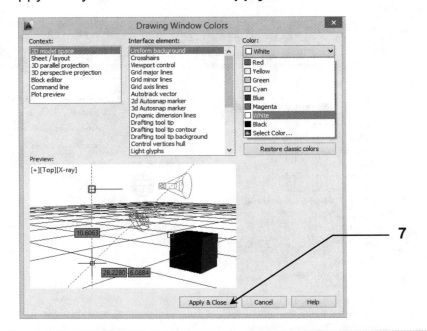

8. Select **Apply** and then **OK** in the **Options** dialog box.

Note: When you change either the color scheme or the drawing area color, the settings remain in the system regardless of which drawing file you use.

Changing the Drawing View

Also during the lessons and exercises you will be asked to change the drawing view to **SE Isometric** (the **SE** means South East). The default drawing view is **Top**. To change the drawing view do the following.

1. Left click on the word "**Top**" in the top left-hand corner of the main drawing area.

2. Select **SE Isometric** from the list.

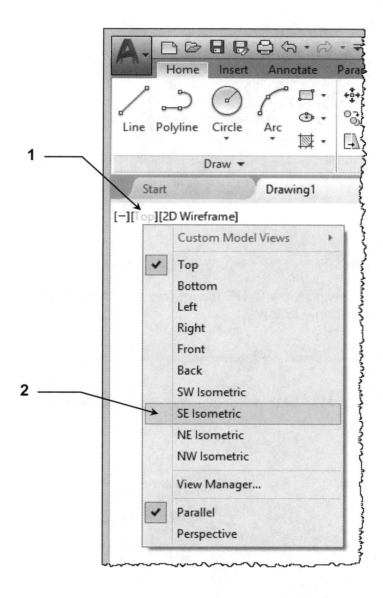

You may use any of the drawing views but the **SE Isometric** view is the one used most in the lessons and exercises throughout this Workbook.

Continued on the next page...

Changing the Drawing View….continued

The example below shows a 3D solid model using the **SE Isometric** view.

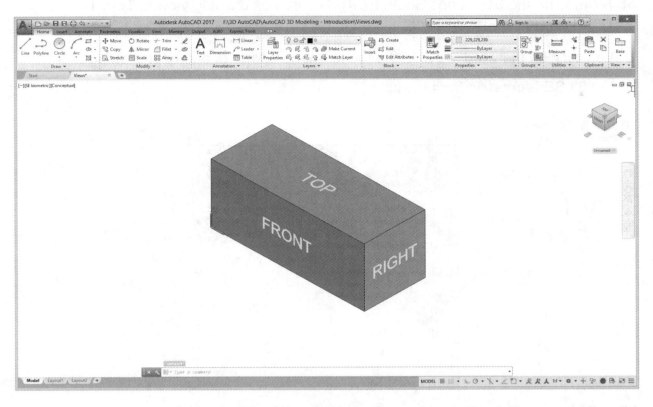

If you were to change the drawing view to **NE Isometric** (North East) view your 3D solid model would look like the example below.

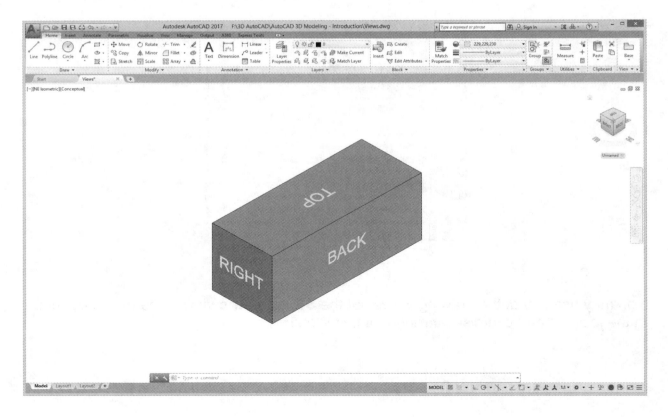

Changing the Visual Style

As well as changing the drawing view in the lessons and exercises, you will also be asked to change the Visual Style to **Conceptual** style. The default Visual Style is **2D Wireframe**. To change the Visual Style do the following.

1. Left click on the word "**2D Wireframe**" in the top left-hand corner of the main drawing area.

2. Select **Conceptual** from the list.

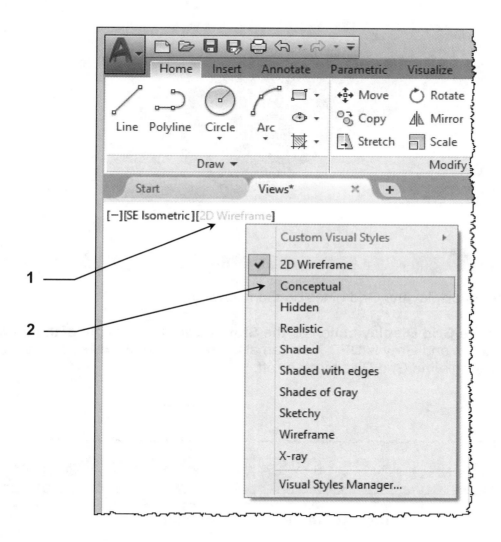

You may wish to experiment with different Visual Styles such as **Realistic** style.

Note: You will notice that when you switch to **Conceptual** style, the main drawing area will revert back to a black color. You will need to open the **Options** dialog box and change the color for **3D parallel projection**, similar to what you did for the 2D Model space.

Continued on the next page...

Changing the Visual Style....continued

You will also notice that in the main drawing area it displays **Grid Lines**.

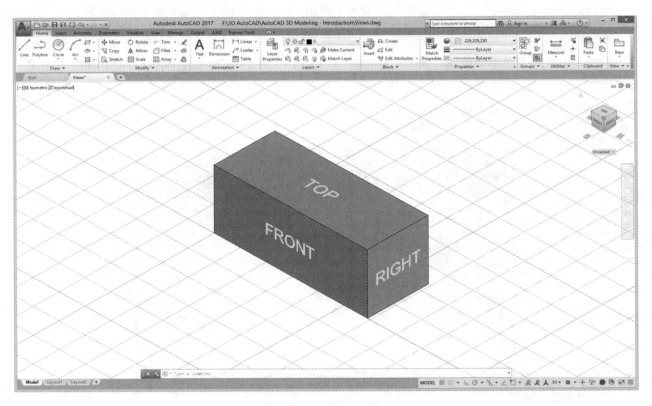

If you prefer not to display the Grid Lines do the following.

1. Left click on the **Grid Display** button on the **Status Bar** at the bottom of the drawing area. Blue is **ON** and Grey is **OFF**. You can also use the **Function** key *F7* on your keyboard to toggle the Grid Display on or off.

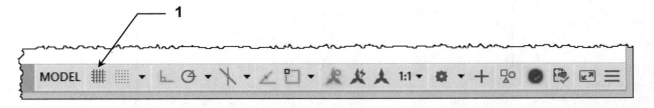

AutoCAD versions 2015 or later.

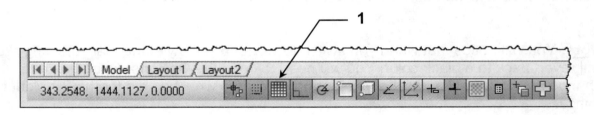

AutoCAD versions 2014 or earlier.

Status Bar Buttons

The **Status Bar** is located at the bottom of the drawing area and contains commonly used drawing tools. You can turn each tool off individually by clicking on the required button or by using the corresponding **Function** key on your keyboard such as *F8* to turn on or off **Ortho Mode** (Ortho is short for Orthographic). A brief description of the Status Bar tools used throughout this Workbook is given below.

Ortho Mode (You may also use the *F8* key to toggle Ortho Mode **ON** or **OFF**.)

Ortho Mode restricts the movement of the cursor to Horizontal or Vertical. When Ortho Mode is **ON** the cursor moves only in the horizontal or vertical directions. When Ortho Mode is **OFF** the cursor moves freely in any direction.

Object Snap (You may also use the *F3* key to toggle 2D Object Snap **ON** or **OFF**.)

When the 2D Object Snap is **ON** the cursor will "snap" to preset locations on 2D objects. You will learn more about Object Snaps in Lesson 4.

AutoCAD versions 2015 or later.

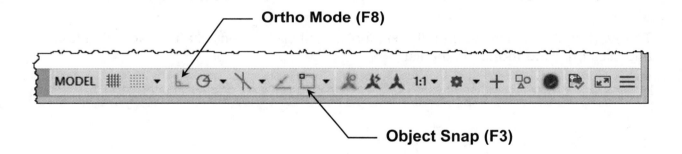

AutoCAD versions 2014 or earlier.

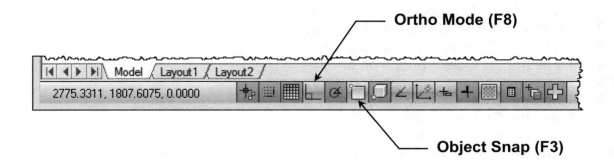

Continued on the next page...

Status Bar Buttons....continued

If you have AutoCAD versions 2014 or earlier you can change the Status Bar tool buttons from icons to text labels. This option is not available for AutoCAD versions 2015 or later.

To change the Status Bar tool buttons to text labels.

1. Right click on any Status Bar tool button and then uncheck **"Use Icons"** in the list.

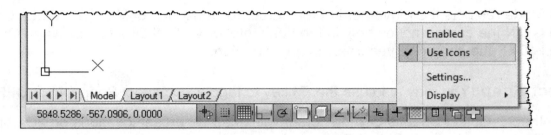

The Status Bar tool buttons will now show as text labels.

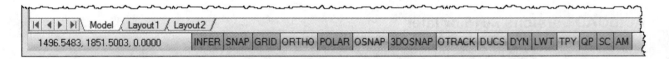

To revert back to icons instead of text labels, right click on any Status Bar tool button and click on **"Use Icons"** in the list.

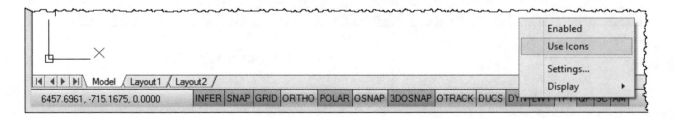

Keyboard Function keys and what they are used for.

F1 - Opens the Help window.
F2 - Displays an Extended Command History list.
F3 - Turns the 2D Object Snaps on or off.
F4 - Turns the 3D Object Snaps on or off.
F5 - Toggles Isoplanes between Top, Right or Left.
F6 - Turns the Dynamic UCS on or off.
F7 - Turns the Grid on or off.
F8 - Turns Orthographic Mode on or off.
F9 - Turns Snap Mode on or off.
F10 - Turns Polar Tracking on or off.
F11 - Turns Object Snap Tracking on or off.
F12 - Turns Dynamic Input on or off.

Using a Wheel Mouse

Using a wheel mouse is by far the easiest and most efficient method of navigating your way around AutoCAD and for creating 3D solid models. The standard wheel mouse has left and right buttons and a central wheel that also doubles as a button. The default functions for the buttons and wheel are as follows.

The **right-hand** button is used to **Enter** a command or for the shortcut **Menu**.

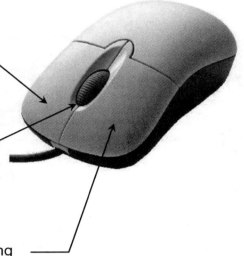

The **Wheel** is used either for **Zooming** and **Panning**, or it can be used for **Zooming** and displaying the **Object Snap Menu**. The default is Zooming and Panning.

The **left-hand** button is used for **Input** during commands and cannot be reprogrammed.

Zooming and Panning with the Wheel.

Zoom

Rotate your Wheel **forward** to Zoom in.

Rotate your Wheel **backward** to Zoom out.

Pan

Press the Wheel and drag the mouse to move the drawing around the screen.

In addition to Zooming and Panning with the Wheel, you can also view the entire drawing by double clicking the Wheel. This is called **Zoom Extents**.

Application and Workspace Descriptions

A Workspace controls the display of Ribbons, Tabs, Menus, Toolbars and Palettes. There are 3 Workspaces to choose from: **2D Drafting & Annotation**, **3D Basics** and **3D Modeling**. Throughout this Workbook you will be using the 2D Drafting & Annotation and 3D Modeling Workspaces.

The default Workspace is **2D Drafting & Annotation**. A brief description of the Workspace elements are shown below. You will learn how to change Workspaces in Lesson 1.

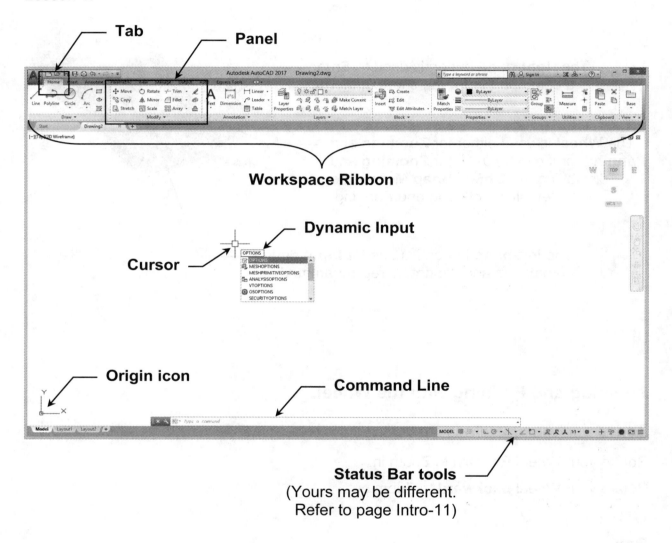

Enlarged view of the **Draw Panel** on the **Home Tab**.

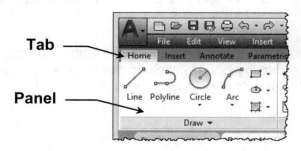

LESSONS

LEARNING OBJECTIVES

After completing this lesson you will be able to:

1. Select a basic 3D Tool.
2. Create a Solid Box.
3. Create a Solid Cylinder.
4. Create a Solid Cone.
5. Create a Solid Sphere.
6. Create a Solid Pyramid.
7. Create a Solid Wedge.
8. Create a Solid Torus.

LESSON 1

Selecting a Basic 3D Tool

In this lesson you will learn how to create 7 basic 3D solid shapes. There are various methods you can use to initiate the commands. The choice is entirely yours regarding which method you choose.

Method 1 – Modeling Panel on the 3D Tools Tab on the Drafting and Annotation Workspace.

By default the **3D Tools Tab** is hidden on the Drafting and Annotation Workspace. To show the **3D Tools Tab** do the following.

1. Right click on any **Panel** of the Workspace Ribbon. For example, right click on the **Draw Panel** of the **Home Tab**.
2. Select **Show Tabs** from the list.
3. Select **3D Tools** from the list.
4. Select the **3D Tools Tab**.
5. Select the **Box** drop-down arrow on the **Modeling Panel** of the **3D Tools Tab**.
6. Select a 3D Tool from the list. For example, **Box**.

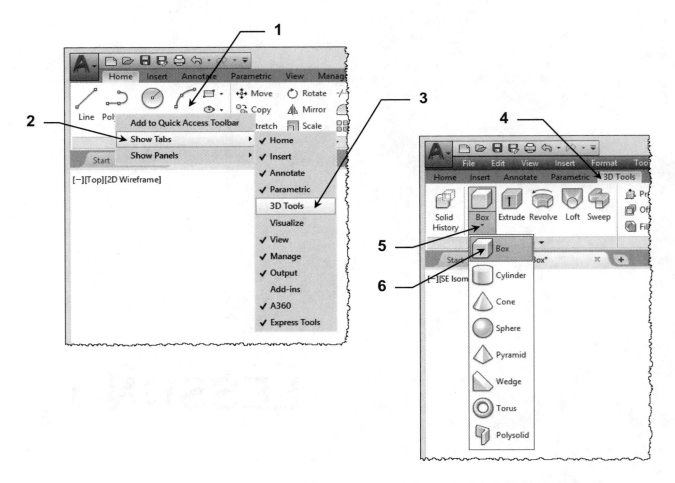

Continued on the next page...

Selecting a Basic 3D Tool….continued

Method 2 – Modeling Panel of the Home Tab on the 3D Modeling Workspace.

The **Drafting and Annotation Workspace** is the default Workspace in AutoCAD. To enable the **3D Modeling Workspace** do the following.

1. Select the **Workspace Switching** icon on the **Status Bar** in the bottom right-hand corner for AutoCAD version 2015 or later. (**Note:** Earlier versions of AutoCAD will be different. The **Workspace Switching** is located on the **Quick Access Toolbar** in the top left-hand corner. See images below.)

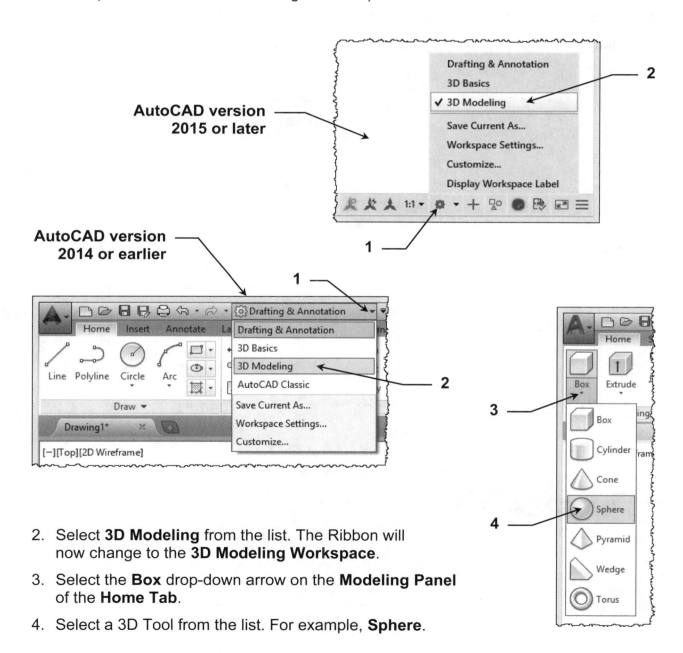

2. Select **3D Modeling** from the list. The Ribbon will now change to the **3D Modeling Workspace**.

3. Select the **Box** drop-down arrow on the **Modeling Panel** of the **Home Tab**.

4. Select a 3D Tool from the list. For example, **Sphere**.

Continued on the next page...

Selecting a Basic 3D Tool....continued

Method 3 – Draw pull-down Menu on the Menu Bar.

The **Menu Bar** is hidden by default on all Workspaces in AutoCAD. To enable the **Menu Bar** do the following.

1. Select the drop-down arrow on the **Quick Access Toolbar**.

2. Select **Show Menu Bar** from the list. The **Menu Bar** will now appear across the top of the Workspace Ribbon.

3. Select the **Draw** pull-down Menu.

4. Select **Modeling** from the list.

5. Select a 3D Tool from the list. For example, **Pyramid**.

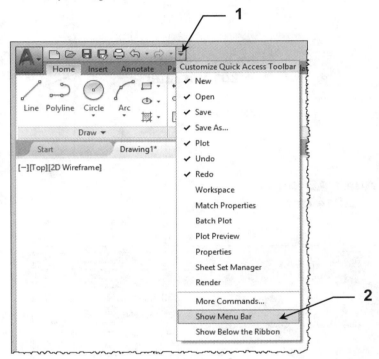

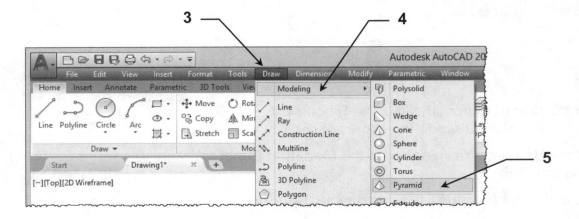

Continued on the next page...

Selecting a Basic 3D Tool....continued

Method 4 – Keyboard entry.

Method 4 is probably the easiest way to initiate a 3D Modeling command if you prefer to use keyboard entry. You can use keyboard entry for all commands within AutoCAD. To initiate a 3D modeling command using keyboard entry do the following.

1. On the **Command Line** or in the **Dynamic Input Box**, type in the name of the 3D Modeling tool you require. For example, *wedge*. (Lowercase is OK.)

2. Press **<enter>** on your keyboard.

1. Command Line

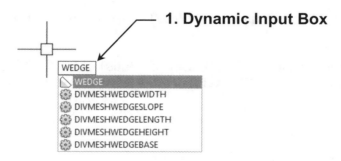

1. Dynamic Input Box

Note: The **Dynamic Input Box** will appear when you start to type in the main drawing area. Make sure Dynamic Input is turned **ON** for this lesson. Pressing **F12** on your keyboard toggles Dynamic Input on or off.

Creating a Solid Box

There are several methods you can use to create a **Solid Box**. The information you have for the size and position of the Box will determine the method you use to create it.

Note: The sizes shown in brackets **[...]** are for metric users. Enter the numbers without the brackets. For example, [12.7] just enter 12.7

Method 1 – Using a start position, diagonal corner position and height of the Box.

1. Start a new drawing file by selecting either **acad.dwt** for inch users, or **acadiso.dwt** for metric users. (Refer to page Intro-2.)

2. Select the **SE Isometric** view. (Refer to page Intro-7.)

3. Select the **Box** tool. (Refer to pages 1-2 through 1-5.)

4. Specify first corner or [Center]: *type in 0,0,0 on your keyboard then press <enter> (P1)*. (You can also place the cursor in any position and then left click with your mouse.)

5. Move your mouse down and to the right. (For X+ and Y+ positions.)

6. Specify other corner or [Cube/Length]: *type in 6,4 [152.4,101.6] then press <enter> (P2)*.

7. Move your mouse up. (For Z+ position.)

8. Specify height or [**2**Point]: *type in 3 [76.2] then press <enter> (P3)*.

A solid Box has been created with a length (X+ Axis) of 6" [152.4 mm], a width (Y+ Axis) of 4" [101.6 mm], and a height (Z+ Axis) of 3" [76.2 mm]. And with a start position of X0, Y0, Z0.

Note: The solid Box is displayed in Conceptual style. (Refer to page Intro-9.)

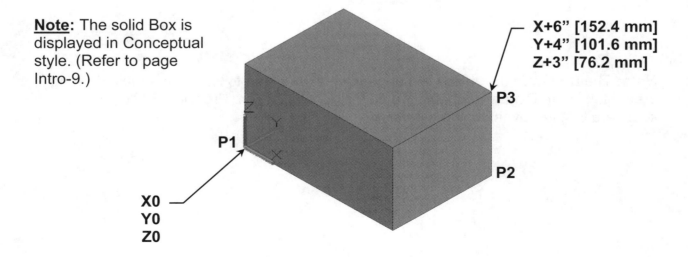

X+6" [152.4 mm]
Y+4" [101.6 mm]
Z+3" [76.2 mm]

P3

P1

P2

X0
Y0
Z0

Creating a Solid Box....continued

Method 2 – Using a start position, length, width and height of the Box.

1. Start a new drawing file by selecting either **acad.dwt** for inch users, or **acadiso.dwt** for metric users.

2. Select the **SE Isometric** view.

3. Select the **Box** tool.

4. Specify first corner or [Center]: *type in 0,0,0 on your keyboard then press <enter> (P1)*. (You can also place the cursor in any position and then left click with your mouse.)

5. Move your mouse up and to the left. (For X– and Y– positions.)

6. Specify other corner or [Cube/Length]: *type in L then press <enter>*.

Note: When using the Length method, make sure **Ortho Mode** is turned on by pressing **F8** on your keyboard or by selecting the **Ortho** icon on the **Status Bar**. This constrains the cursor to just horizontal and vertical movements. (Refer to page Intro-11.)

7. Specify length: *type in 5.25 [133.35] then press <enter>*.

8. Specify width: *type in 3.2 [81.28] then press <enter>*.

9. Move your mouse down. (For Z– position.)

10. Specify height or [**2P**oint]: *type in 2.5 [63.5] then press <enter>*.

A solid Box has been created with a length (X– Axis) of 5.25" [133.35 mm], a width (Y– Axis) of 3.2" [81.28 mm], and a height (Z– Axis) of 2.5" [63.5 mm]. And with a start position of X0, Y0, Z0.

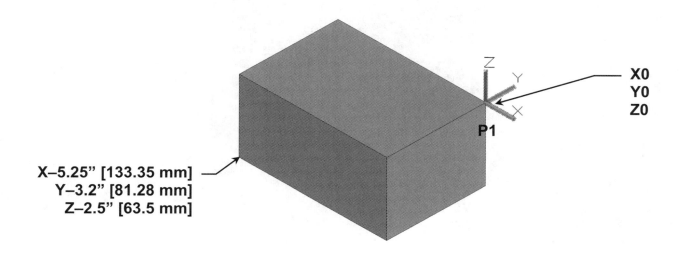

X–5.25" [133.35 mm]
Y–3.2" [81.28 mm]
Z–2.5" [63.5 mm]

X0
Y0
Z0

P1

Creating a Solid Box....continued

Method 3 – Using a start position, with the length, width and height of the Box all having the same dimensions.

1. Start a new drawing file by selecting either **acad.dwt** for inch users, or **acadiso.dwt** for metric users.

2. Select the **SE Isometric** view.

3. Select the **Box** tool.

4. Specify first corner or [Center]: *type in 0,0,0 on your keyboard then press <enter> (P1)*. (You can also place the cursor in any position and then left click with your mouse.)

5. Move your mouse up and to the right. (For X– , Y+ and Z+ positions.)

6. Specify other corner or [Cube/Length]: *type in C then press <enter>*.

7. Specify length: *type in 5 [127] then press <enter>*.

A solid Box has been created with a length (X– Axis) of 5" [127 mm], a width (Y+ Axis) of 5" [127 mm], and a height (Z+ Axis) of 5" [127 mm]. And with a start position of X0, Y0, Z0.

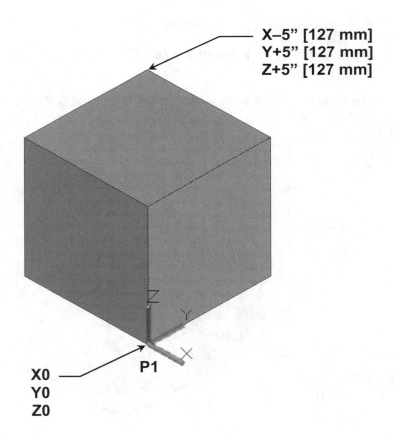

X–5" [127 mm]
Y+5" [127 mm]
Z+5" [127 mm]

P1

X0
Y0
Z0

Creating a Solid Box....continued

Method 4 – Using the center position of the Box, a corner position and height.

1. Start a new drawing file by selecting either **acad.dwt** for inch users, or **acadiso.dwt** for metric users.

2. Select the **SE Isometric** view.

3. Select the **Box** tool.

4. Specify first corner or [Center]: *type in C then press <enter>.*

5. Specify center: *type in 0,0,0 on your keyboard then press <enter> (P1).* (You can also place the cursor in any position and then left click with your mouse.)

6. Move your mouse to the right.

7. Specify corner or [Cube/Length]: *type in 3,3,1 [76.2,76.2,25.4] then press <enter>.*

A solid Box has been created with an overall length of 6" [152.4 mm], an overall width of 6" [152.4 mm], and an overall height of 2" [50.8 mm]. And with a start position of X0, Y0, Z0 at the center of the Box.

Note: Ortho Mode is automatically turned off using this method. It turns back on after the command is complete.

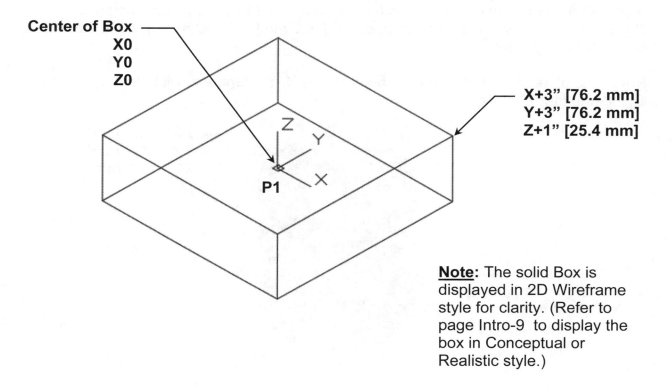

Center of Box
X0
Y0
Z0

X+3" [76.2 mm]
Y+3" [76.2 mm]
Z+1" [25.4 mm]

P1

Note: The solid Box is displayed in 2D Wireframe style for clarity. (Refer to page Intro-9 to display the box in Conceptual or Realistic style.)

Creating a Solid Cylinder

There are various methods you can use to create a **Solid Cylinder**. The information you have for the size and position of the Cylinder will determine the method you use to create it.

Note: The sizes shown in brackets **[...]** are for metric users. Enter the numbers without the brackets. For example, [12.7] just enter 12.7

Method 1 – Using the base center position of the Cylinder, a radius and height.

1. Start a new drawing file by selecting either **acad.dwt** for inch users, or **acadiso.dwt** for metric users.

2. Select the **SE Isometric** view.

3. Select the **Cylinder** tool.

4. Specify center point of base or [3P/2P/Ttr/Elliptical]: *type in 3,2,0 [76.2,50.8,0] on your keyboard then press <enter> (P1)*. (You can also place the cursor in any position and then left click with your mouse.)

5. Specify base radius or [Diameter]: *type in 1.5 [38.1] then press <enter>*.

6. Move your mouse up. (For Z+ position.)

7. Specify height or [2Point/Axis endpoint]: *type in 3 [76.2] then press <enter> (P2)*.

A solid Cylinder has been created with a radius of 1.5" [38.1 mm], and a height of 3" [76.2 mm]. And with a start position of X3" [76.2 mm], Y2" [50.8 mm], Z0 at the base center of the Cylinder.

Note: The base of the Cylinder is located on the "**X, Y**" plane and with the height of the Cylinder in the "**Z**" plane.

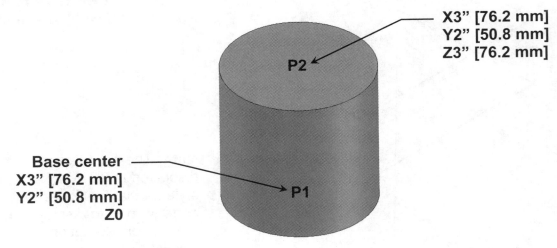

X3" [76.2 mm]
Y2" [50.8 mm]
Z3" [76.2 mm]

P2

Base center
X3" [76.2 mm]
Y2" [50.8 mm]
Z0

P1

Creating a Solid Cylinder....continued

Method 2 – Using the base center position of the Cylinder, a diameter and height.

1. Start a new drawing file by selecting either **acad.dwt** for inch users, or **acadiso.dwt** for metric users.

2. Select the **SE Isometric** view.

3. Select the **Cylinder** tool.

4. Specify center point of base or [**3P/2P/T**tr/**E**lliptical]: *type in –1,–2,0 [–25.4,–50.8,0] on your keyboard then press <enter> (P1)*. (You can also place the cursor in any position and then left click with your mouse.)

5. Specify base radius or [Diameter]: *type in D then press <enter>*.

6. Specify diameter: *type in 2.8 [71.12] then press <enter>*.

7. Move your mouse up. (For Z+ position.)

8. Specify height or [**2P**oint/**A**xis endpoint]: *type in 4.25 [107.95] then press <enter> (P2)*.

A solid Cylinder has been created with a diameter of 2.8" [71.12 mm], and a height of 4.25" [107.95 mm]. And with a start position of X–1" [–25.4 mm], Y–2" [–50.8 mm], Z0 at the base center of the Cylinder.

Note: The base of the Cylinder is located on the "**X, Y**" plane and with the height of the Cylinder in the "**Z**" axis.

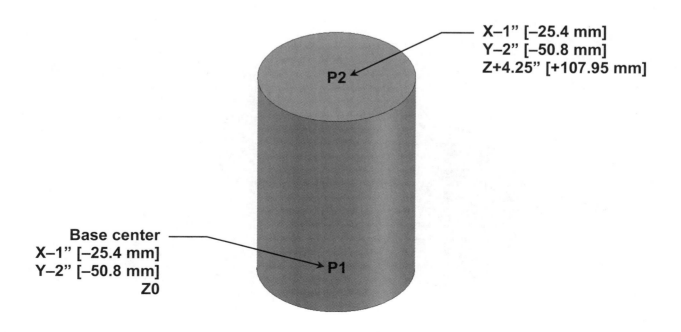

X–1" [–25.4 mm]
Y–2" [–50.8 mm]
Z+4.25" [+107.95 mm]

P2

Base center
X–1" [–25.4 mm]
Y–2" [–50.8 mm]
Z0

P1

Creating a Solid Cylinder....continued

Method 3 – Using the base center position of the Cylinder, a radius and axis endpoint.

1. Start a new drawing file by selecting either **acad.dwt** for inch users, or **acadiso.dwt** for metric users.

2. Select the **SE Isometric** view.

3. Select the **Cylinder** tool.

4. Specify center point of base or [3P/2P/Ttr/Elliptical]: *type in 0,0,0 on your keyboard then press <enter> (P1)*. (You can also place the cursor in any position and then left click with your mouse.)

5. Specify base radius or [Diameter]: *type in 1.181 [30] then press <enter>*.

6. Specify height or [2Point/Axis endpoint]: *type in A then press <enter>*.

7. Turn on **Ortho Mode (F8)** then move your mouse up and to the right. (For Y+ position.)

8. Specify axis endpoint: *type in 3.937 [100] then press <enter> (P2)*.

A solid Cylinder has been created with a radius of 1.181" [30 mm], and a length of 3.937" [100 mm]. And with a start position of X0, Y0, Z0 at the base center of the Cylinder.

Note: The base of the Cylinder is located on the "**X, Z**" plane and with the length of the Cylinder in the "**Y**" axis.

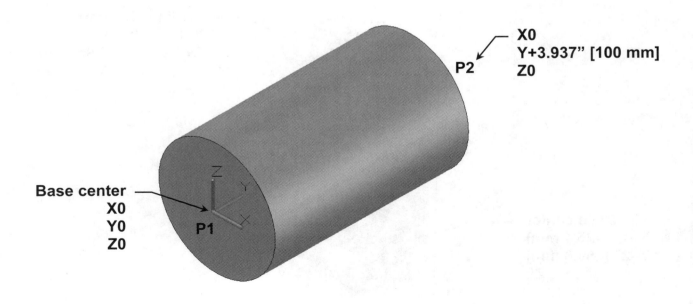

Creating a Solid Cone

There are 2 methods you can use to create a **Solid Cone**. You can create a Cone with an apex (sharp point), or a truncated Cone, which has a flat top parallel to the base.

Note: The sizes shown in brackets **[...]** are for metric users. Enter the numbers without the brackets. For example, [12.7] just enter 12.7

Method 1 – Using the base center position of the Cone, a radius and height.

1. Start a new drawing file by selecting either **acad.dwt** for inch users, or **acadiso.dwt** for metric users.

2. Select the **SE Isometric** view.

3. Select the **Cone** tool.

4. Specify center point of base or [**3P/2P/T**tr/**E**lliptical]: *type in 0,0,0 on your keyboard then press <enter> (P1).* (You can also place the cursor in any position and then left click with your mouse.)

5. Specify base radius or [Diameter]: *type in 1.125 [28.57] then press <enter>.*

6. Move your mouse up. (For Z+ position.)

7. Specify height or [**2P**oint/**A**xis endpoint/**T**op radius]: *type in 2.75 [69.85] then press <enter> (P2).*

A solid Cone has been created with a radius of 1.125" [28.57 mm], and a height of 2.75" [69.85 mm]. And with a start position of X0, Y0, Z0 at the base center of the Cone.

Note: The base of the Cone is located on the "**X**, **Y**" plane and with the height of the Cone in the "**Z**" axis. (You can also create a Cone using the axis endpoint. Refer to page 1-12.)

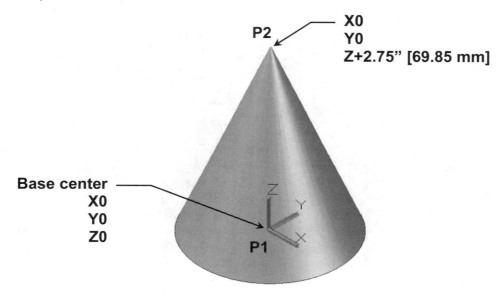

P2 — X0 / Y0 / Z+2.75" [69.85 mm]

Base center X0 Y0 Z0

P1

Creating a Solid Cone....continued

Method 2 – Using the base center position of the Cone, a base radius, a top radius and height.

1. Start a new drawing file by selecting either **acad.dwt** for inch users, or **acadiso.dwt** for metric users.

2. Select the **SE Isometric** view.

3. Select the **Cone** tool.

4. Specify center point of base or [3P/2P/Ttr/Elliptical]: *type in 0,0,0 on your keyboard then press <enter> (P1)*. (You can also place the cursor in any position and then left click with your mouse.)

5. Specify base radius or [Diameter]: *type in 1.5 [38.1] then press <enter>*.

6. Specify height or [2Point/Axis endpoint/Top radius]: *type in T then press <enter>*.

7. Specify top radius: *type in .5 [12.7] then press <enter>*.

8. Move your mouse up. (For Z+ position.)

9. Specify height or [2Point/Axis endpoint]: *type in 2.5 [63.5] then press <enter> (P2)*.

A solid Cone has been created with a base radius of 1.5" [38.1 mm], a top radius of 0.5" [12.7 mm], and a height of 2.5" [63.5 mm]. And with a start position of X0, Y0, Z0 at the base center of the Cone.

Note: The base of the Cone is located on the "**X, Y**" plane and with the height of the Cone in the "**Z**" axis. (You can also create a Cone using the axis endpoint. Refer to page 1-12.)

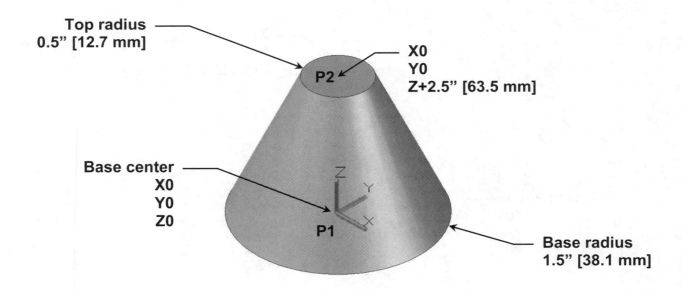

Top radius
0.5" [12.7 mm]

P2

X0
Y0
Z+2.5" [63.5 mm]

Base center
X0
Y0
Z0

P1

Base radius
1.5" [38.1 mm]

Creating a Solid Sphere

A **Solid Sphere** can be created by specifying the center point location and a radius or diameter for the size.

Note: The sizes shown in brackets **[...]** are for metric users. Enter the numbers without the brackets. For example, [12.7] just enter 12.7

Method 1 – Using the center position of the Sphere and radius.

1. Start a new drawing file by selecting either **acad.dwt** for inch users, or **acadiso.dwt** for metric users.

2. Select the **SE Isometric** view.

3. Select the **Sphere** tool.

4. Specify center point or [3P/2P/Ttr]: *type in 0,0,0 on your keyboard then press* *<enter> (P1)*. (You can also place the cursor in any position and then left click with your mouse.)

5. Specify radius or [Diameter]: *type in 1.575 [40] then press <enter>*.

Note: If you want to enter the diameter of the Sphere type in **D** then press **<enter>**, then type in the diameter and press **<enter>** again.

A solid Sphere has been created with a radius of 1.575" [40 mm]. And with a start position of X0, Y0, Z0 at the center of the Sphere.

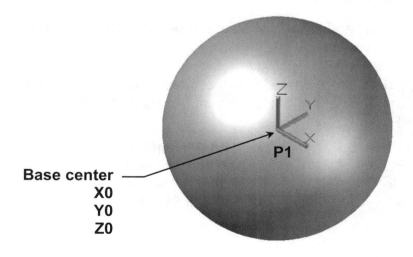

Base center
X0
Y0
Z0

P1

Creating a Solid Pyramid

There are various methods you can use to create a **Solid Pyramid**. You can create a Pyramid that has the base circumscribed around a radius, or you can have the base inscribed within a radius. You can also have the Pyramid as an apex (sharp point), or a truncated Pyramid, which has a flat top parallel to the base. You can choose to have between 3 and 32 sides for the pyramid.

Note: The sizes shown in brackets [...] are for metric users. Enter the numbers without the brackets. For example, [12.7] just enter 12.7

Method 1 – Using the center position of the Pyramid with the base circumscribed around a radius.

1. Start a new drawing file by selecting either **acad.dwt** for inch users, or **acadiso.dwt** for metric users.

2. Select the **SE Isometric** view.

3. Select the **Pyramid** tool.

4. Specify center point of base or [Edge/Sides]: *type in S on your keyboard then press <enter>.*

5. Enter number of sides [4]: *type in 6 then press <enter>.*

6. Specify center point of base or [Edge/Sides]: *type in 0,0,0 then press <enter> (P1).*

7. Specify base radius or [Inscribed]: *type in 1.4 [35.56] then press <enter>.*

8. Move your mouse up. (For Z+ position.)

9. Specify height or [2Point/Axis endpoint/Top radius]: *type in 4 [101.6] then press <enter> (P2).*

A solid Pyramid with 6 sides has been created with the base circumscribed around a radius of 1.4" [35.56 mm], and a height of 4" [101.6 mm]. And with a start position of X0, Y0, Z0 at the center of the Pyramid.

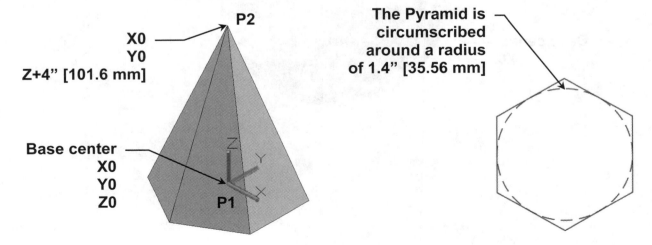

P2

X0
Y0
Z+4" [101.6 mm]

The Pyramid is circumscribed around a radius of 1.4" [35.56 mm]

Base center
X0
Y0
Z0

P1

Creating a Solid Pyramid....continued

Method 2 – Using the center position of the Pyramid with the base inscribed within a radius.

1. Start a new drawing file by selecting either **acad.dwt** for inch users, or **acadiso.dwt** for metric users.

2. Select the **SE Isometric** view.

3. Select the **Pyramid** tool.

4. Specify center point of base or [Edge/Sides]: *type in S on your keyboard then press <enter>*.

5. Enter number of sides [4]: *type in 6 then press <enter>*.

6. Specify center point of base or [Edge/Sides]: *type in 0,0,0 then press <enter> (P1)*.

7. Specify base radius or [Inscribed]: *type in I (for Inscribed) then press <enter>*.

8. Specify base radius or [Circumscribed]: *type in 1.4 [35.56] then press <enter>*.

9. Move your mouse up. (For Z+ position.)

10. Specify height or [2Point/Axis endpoint/Top radius]: *type in 4 [101.6] then press <enter> (P2)*.

A solid Pyramid with 6 sides has been created with the base inscribed within a radius of 1.4" [35.56 mm], and a height of 4" [101.6 mm]. And with a start position of X0, Y0, Z0 at the center of the Pyramid.

Note: You can also create a truncated Pyramid, which has a flat top parallel to the base. (Refer to page 1-14.)

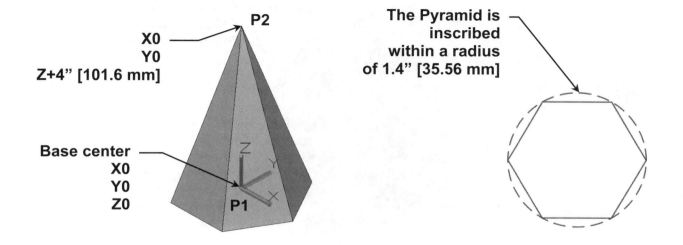

P2

X0
Y0
Z+4" [101.6 mm]

The Pyramid is inscribed within a radius of 1.4" [35.56 mm]

Base center
X0
Y0
Z0

P1

Creating a Solid Wedge

There are various methods you can use to create a **Solid Wedge**. The information you have for the size and position of the Wedge will determine the method you use to create it.

Note: The sizes shown in brackets **[...]** are for metric users. Enter the numbers without the brackets. For example, [12.7] just enter 12.7

Method 1 – Using a start position, diagonal corner position and height of the Wedge.

1. Start a new drawing file by selecting either **acad.dwt** for inch users, or **acadiso.dwt** for metric users.

2. Select the **SE Isometric** view.

3. Select the **Wedge** tool.

4. Specify first corner or [Center]: *type in 0,0,0 on your keyboard then press <enter> (P1)*. (You can also place the cursor in any position and then left click with your mouse.)

5. Move your mouse down and to the right. (For X+ and Y+ positions.)

6. Specify other corner or [Cube/Length]: *type in 3.5,2.5 [88.9,63.5] then press <enter> (P2)*. (You can also place the cursor in any position and then left click with your mouse.)

7. Move your mouse up. (For Z+ position.)

8. Specify height or [**2P**oint]: *type in 2.5 [63.5] then press <enter> (P3)*.

A solid Wedge has been created with a length (X+ Axis) of 3.5" [88.9 mm], a width (Y+ Axis) of 2.5" [63.5 mm], and a height (Z+ Axis) of 2.5" [63.5 mm]. And with a start position of X0, Y0, Z0.

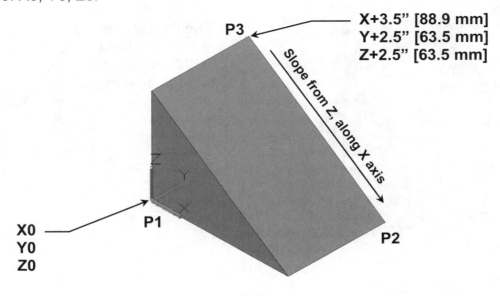

P3

X+3.5" [88.9 mm]
Y+2.5" [63.5 mm]
Z+2.5" [63.5 mm]

Slope from Z, along X axis

X0
Y0
Z0

P1

P2

Creating a Solid Wedge....continued

Method 2 – Using a start position, length, width and height of the Wedge.

1. Start a new drawing file by selecting either **acad.dwt** for inch users, or **acadiso.dwt** for metric users.

2. Select the **SE Isometric** view.

3. Select the **Wedge** tool.

4. Specify first corner or [Center]: *type in 0,0,0 on your keyboard then press <enter> (P1)*. (You can also place the cursor in any position and then left click with your mouse.)

5. Move your mouse down and to the right. (For X+ and Y+ positions.)

6. Specify other corner or [Cube/Length]: *type in L* (for Length) *then press <enter>*.

Note: When using the Length method, make sure **Ortho Mode** is turned on by pressing **F8** on your keyboard or by selecting the **Ortho** icon on the **Status Bar**. This constrains the cursor to just horizontal and vertical movements.

7. Specify length: *type in 3.937 [100] then press <enter>*.

8. Specify width: *type in 1.968 [50] then press <enter>*.

9. Move your mouse up. (For Z+ position.)

10. Specify height or [**2P**oint]: *type in 1.968 [50] then press <enter> (P2)*.

A solid Wedge has been created with a length (X+ Axis) of 3.937" [100 mm], a width (Y+ Axis) of 1.968" [50 mm], and a height (Z+ Axis) of 1.968" [50 mm]. And with a start position of X0, Y0, Z0.

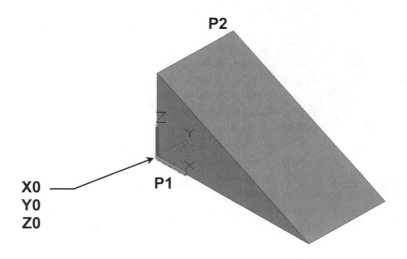

Creating a Solid Wedge....continued

Method 3 – Using a start position, with the length, width and height of the Wedge all having the same dimensions.

1. Start a new drawing file by selecting either **acad.dwt** for inch users, or **acadiso.dwt** for metric users.

2. Select the **SE Isometric** view.

3. Select the **Wedge** tool.

4. Specify first corner or [Center]: *type in 0,0,0 on your keyboard then press <enter> (P1)*. (You can also place the cursor in any position and then left click with your mouse.)

5. Move your mouse to the right. (For X+ , Y+ and Z+ positions.)

6. Specify other corner or [Cube/Length]: *type in C then press <enter>*.

7. Specify length: *type in 2.953 [75] then press <enter>*.

A solid Wedge has been created with a length (X+ Axis) of 2.953" [75 mm], a width (Y+ Axis) of 2.953" [75 mm], and a height (Z+ Axis) of 2.953" [75 mm] And with a start position of X0, Y0, Z0.

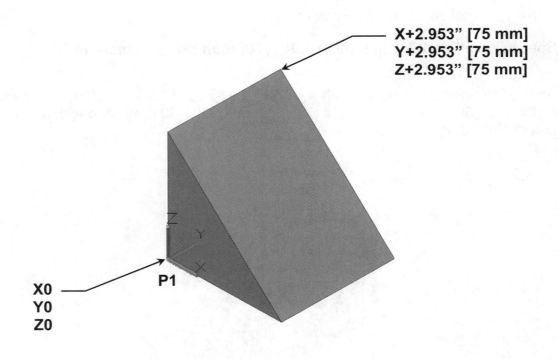

X+2.953" [75 mm]
Y+2.953" [75 mm]
Z+2.953" [75 mm]

P1

X0
Y0
Z0

Creating a Solid Wedge....continued

Method 4 – Using the center position of the Wedge, a corner position and height.

1. Start a new drawing file by selecting either **acad.dwt** for inch users, or **acadiso.dwt** for metric users.

2. Select the **SE Isometric** view.

3. Select the **Wedge** tool.

4. Specify first corner or [Center]: *type in C then press <enter>.*

5. Specify center: *type in 0,0,0 on your keyboard then press <enter> (P1)*. (You can also place the cursor in any position and then left click with your mouse.)

6. Move your mouse to the right.

7. Specify corner or [Cube/Length]: *type in 3,3,1.5 [76.2,76.2,38.1] then press <enter>.*

A solid Wedge has been created with an overall length of 6" [152.4 mm], an overall width of 6" [152.4 mm], and an overall height of 3" [76.2 mm]. And with a start position of X0, Y0, Z0 at the center of the Wedge.

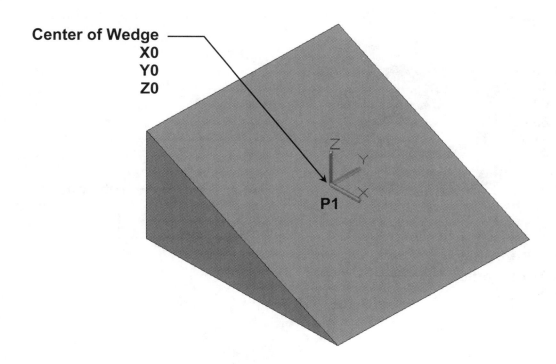

Center of Wedge
X0
Y0
Z0

P1

Creating a Solid Torus

There are 3 shapes you can create using the **Torus** command.

Note: The sizes shown in brackets **[...]** are for metric users. Enter the numbers without the brackets. For example, [12.7] just enter 12.7

Method 1 – Donut shape. The Torus radius is greater than the tube radius.

1. Start a new drawing file by selecting either **acad.dwt** for inch users, or **acadiso.dwt** for metric users.

2. Select the **SE Isometric** view.

3. Select the **Torus** tool.

4. Specify center point or [**3P/2P/T**tr]: *type in 0,0,0 on your keyboard then press <enter> (P1).* (You can also place the cursor in any position and then left click with your mouse.)

5. Specify radius or [Diameter]: *type in 1.5 [38.1] on your keyboard then press <enter>*.

6. Specify tube radius or [**2P**oint/**D**iameter]: *type in .5 [12.7] then press <enter>*.

A solid Torus has been created with an overall radius of 1.5" [38.1 mm], and a tube radius of 0.5" [12.7 mm]. And with a start position of X0, Y0, Z0 at the center of the Torus.

Note: It is important to remember that the Torus radius must be greater than the tube radius.

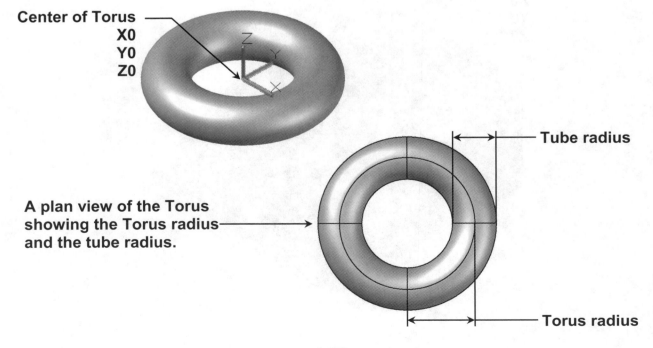

Center of Torus
X0
Y0
Z0

A plan view of the Torus showing the Torus radius and the tube radius.

Tube radius

Torus radius

Creating a Solid Torus....continued

Method 2 – Football shape. The Torus radius must be negative with the tube radius positive and greater than the Torus radius.

1. Start a new drawing file by selecting either **acad.dwt** for inch users, or **acadiso.dwt** for metric users.

2. Select the **SE Isometric** view.

3. Select the **Torus** tool.

4. Specify center point or [3P/2P/Ttr]: *type in 0,0,0 on your keyboard then press <enter> (P1).* (You can also place the cursor in any position and then left click with your mouse.)

5. Specify radius or [Diameter]: *type in –3 [–76.2] on your keyboard then press <enter>.*

6. Specify tube radius or [2Point/Diameter]: *type in 5 [127] then press <enter>.*

A solid Torus has been created with a Torus radius of –3" [–76.2 mm], and a tube radius of 5" [127 mm]. And with a start position of X0, Y0, Z0 at the center of the Torus.

Note: It is important to remember that the Torus radius must be negative with the tube radius positive and greater than the Torus radius.

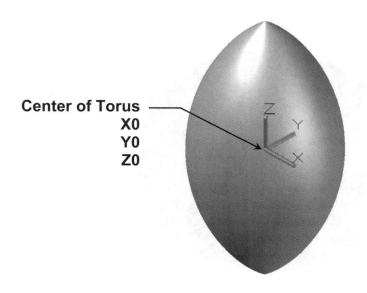

Center of Torus
X0
Y0
Z0

Creating a Solid Torus....continued

Method 3 – Self-intersecting. The Torus radius must be less than the tube radius.

1. Start a new drawing file by selecting either **acad.dwt** for inch users, or **acadiso.dwt** for metric users.

2. Select the **SE Isometric** view.

3. Select the **Torus** tool.

4. Specify center point or [3P/2P/Ttr]: *type in 0,0,0 on your keyboard then press <enter> (P1).* (You can also place the cursor in any position and then left click with your mouse.)

5. Specify radius or [Diameter]: *type in 1 [25.4] on your keyboard then press <enter>.*

6. Specify tube radius or [2Point/Diameter]: *type in 2.5 [63.5] then press <enter>.*

A solid Torus has been created with a Torus radius of 1" [25.4 mm], and a tube radius of 2.5" [63.5 mm]. And with a start position of X0, Y0, Z0 at the center of the Torus.

Note: It is important to remember that the Torus radius must be less than the tube radius.

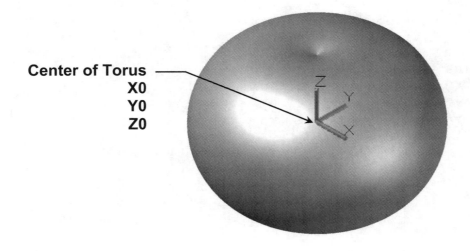

Center of Torus
X0
Y0
Z0

Exercise 1A

1. Start a new drawing file by selecting either **acad.dwt** for inch users, or **acadiso.dwt** for metric users.

2. Select the **SE Isometric** view.

3. Save the drawing file as **Ex-1A** but keep the drawing file open.

4. Use the **Box** tool and **Method 1** to create 4 solid boxes, using a start position, diagonal corner position and a height. Make sure **Dynamic Input** is turned on (*F12*).

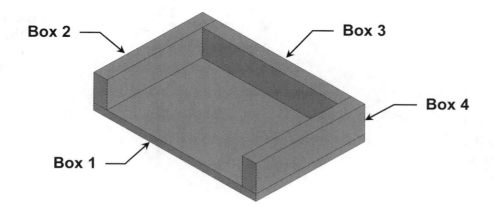

Box 2 —   — Box 3

Box 4

Box 1 —

Box 1

Inch users

Start position = 0,0,0
Diagonal corner position = 3,2
Height = 0.125

Metric users

Start position = 0,0,0
Diagonal corner position = 76.2,50.8
Height = 3.17

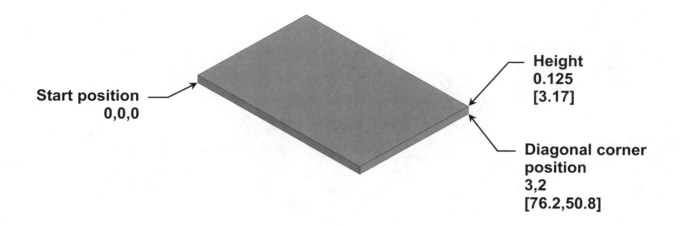

Start position
0,0,0

Height
0.125
[3.17]

Diagonal corner
position
3,2
[76.2,50.8]

Continued on the next page...

Exercise 1A....continued

Box 2

Inch users

Start position = 0,0,0.125
Diagonal corner position = 0.25,2
Height = 0.5

Metric users

Start position = 0,0,3.17
Diagonal corner position = 6.35,50.8
Height = 12.7

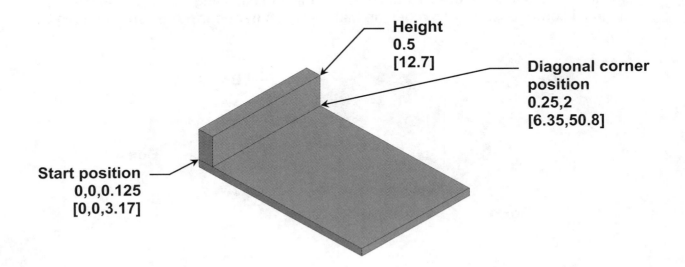

**Height
0.5
[12.7]**

**Diagonal corner
position
0.25,2
[6.35,50.8]**

**Start position
0,0,0.125
[0,0,3.17]**

Box 3

Inch users

Start position = 0.25,1.75,0.125
Diagonal corner position = 2.5,0.25
Height = 0.500

Metric users

Start position = 6.35,44.45,3.17
Diagonal corner position = 63.5,6.35
Height = 12.7

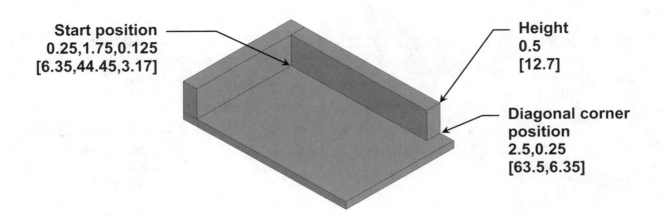

**Start position
0.25,1.75,0.125
[6.35,44.45,3.17]**

**Height
0.5
[12.7]**

**Diagonal corner
position
2.5,0.25
[63.5,6.35]**

Continued on the next page...

Exercise 1A....continued

Box 4

Inch users

Start position = 2.75,0,0.125
Diagonal corner position = 0.25,2
Height = 0.500

Metric users

Start position = 69.85,0,3.17
Diagonal corner position = 6.35,50.8
Height = 12.7

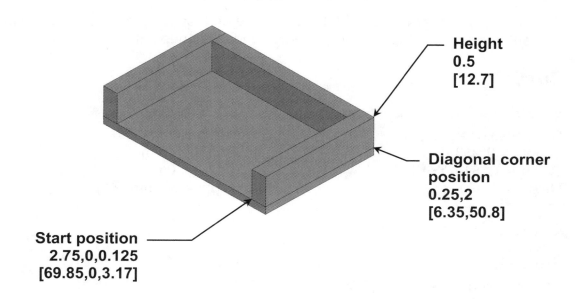

**Height
0.5
[12.7]**

**Diagonal corner
position
0.25,2
[6.35,50.8]**

**Start position
2.75,0,0.125
[69.85,0,3.17]**

5. Save the drawing file again as **Ex-1A** (Overwrite the existing one).

Note: When creating the 4 boxes, the diagonal corner position is **Relative** to the start position (the last point you entered). For example, the start position for Box 4 was 2.75,0,0.125 [69.85,0,3.17]; this is the **Absolute** position from 0,0,0. The diagonal corner position was 0.25,2 [6.35,50.8]; this is the **Relative** position from the start position, the actual base size of the box.

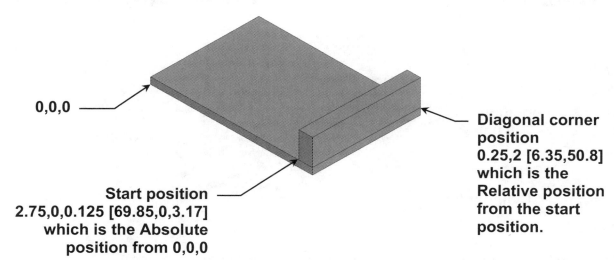

0,0,0

**Diagonal corner
position
0.25,2 [6.35,50.8]
which is the
Relative position
from the start
position.**

**Start position
2.75,0,0.125 [69.85,0,3.17]
which is the Absolute
position from 0,0,0**

Exercise 1B

1. Start a new drawing file by selecting either **acad.dwt** for inch users, or **acadiso.dwt** for metric users.

2. Select the **SE Isometric** view.

3. Save the drawing file as **Ex-1B** but keep the drawing file open.

4. Use the **Box** tool and **Method 2** to create a solid box, using a start position, length, width and a height. Make sure **Dynamic Input** is turned on (*F12*).

5. Use the **Cylinder** tool and **Method 3** to create 2 solid cylinders, using the base center position of the cylinder, a radius and axis endpoint.

Box

Inch users

Start position = 0,0,0
Length = 2.953" (X+)
Width = 1.181" (Y+)
Height = 1.181" (Z+)

Metric users

Start position = 0,0,0
Length = 75 mm (X+)
Width = 30 mm (Y+)
Height = 30 mm (Z+)

Cylinder 1

Inch users

Start position = 2.953,0.59,0.59
Radius = 0.236
Axis endpoint = 0.984 (X+)

Metric users

Start position = 75,15,15
Radius = 6
Axis endpoint = 25 (X+)

Cylinder 2

Inch users

Start position = 1.476,0,0.59
Radius = 0.236
Axis endpoint = 0.984 (Y–)

Metric users

Start position = 37.5,0,15
Radius = 6
Axis endpoint = 25 (Y–)

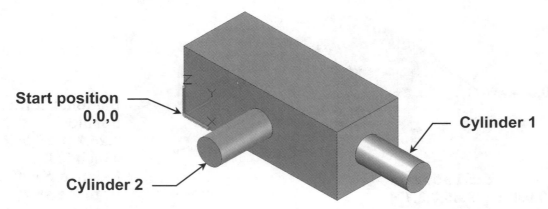

6. Save the drawing file again as **Ex-1B** (overwrite the existing one).

LEARNING OBJECTIVES

After completing this lesson you will be able to:

1. Open the Properties Palette.
2. Modify Solids Using the Properties Palette.
3. Modify Solids Using the Grips.

LESSON 2

Opening the Properties Palette

The **Properties Palette** is one of the easiest methods you can use to modify the sizes of 3D solid models. There are various methods you can use to open the Properties Palette.

Method 1 – Properties Panel on the Home Tab of the Drafting and Annotation Workspace.

1. Left click on a 3D solid model to select it. The model will highlight blue.

2. Select the diagonal arrow on the Properties Panel of the Home Tab.

Method 2 – Palettes Panel on the View Tab of the 3D Modeling Workspace.

1. Left click on a 3D solid model to select it. The model will highlight blue.

2. Select the Properties button on the Palettes Panel of the View Tab.

Method 3 – Ctrl + 1 on your keyboard.

1. Left click on a 3D solid model to select it. The model will highlight blue.

2. Press **Ctrl + 1** on your keyboard.

Opening the Properties Palette....continued

Method 4 – Right click Menu.

1. Left click on a 3D solid model to select it. The model will highlight blue.

2. Right click with your mouse and select **Properties** from the Menu.

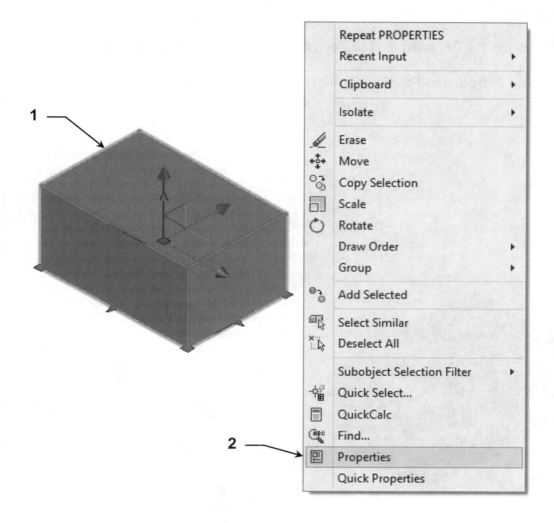

The choice is yours regarding which method you choose to open the Properties Palette. Many users prefer to keep the Properties Palette open all the time. If you wish to close the Palette simply click on the "**X**" or press **Ctrl + 1** on your keyboard.

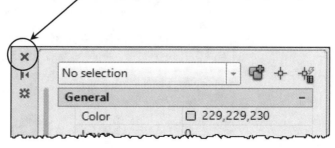

Modify Solids Using the Properties Palette

An example of modifying the sizes of a 3D solid Box.

1. Start a new drawing file by selecting either **acad.dwt** for inch users, or **acadiso.dwt** for metric users.

2. Create a 3D solid Box with a length of 6" [152.4 mm], a width of 4" [101.6 mm], and a height of 3" [76.2 mm].

3. Left click on the 3D solid Box to select it. The Box will highlight blue.

4. Open the **Properties Palette**.

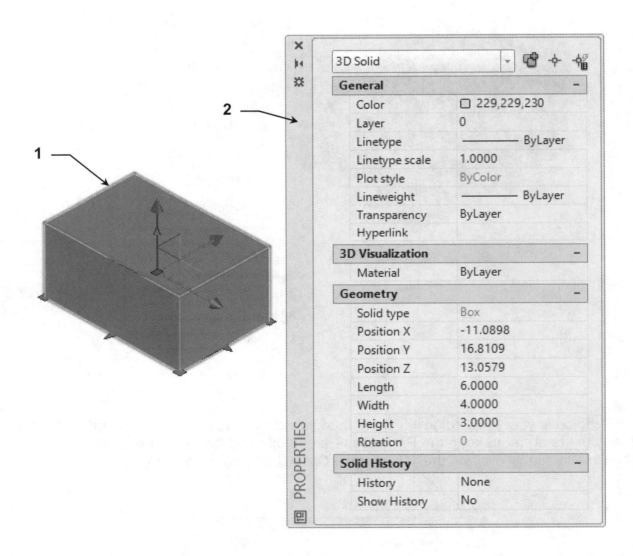

Continued on the next page...

Modify Solids Using the Properties Palette....continued

5. Select the Length dimension (it will highlight blue) and change it to 5.905 [150].

6. Select the Width dimension and change it to 3.937 [100].

7. Select the Height dimension and change it to 0.984 [25] then press **<enter>**.

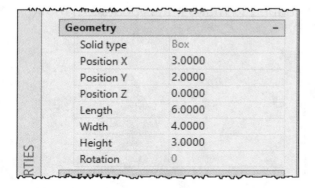

Inch dimensions before modifying.

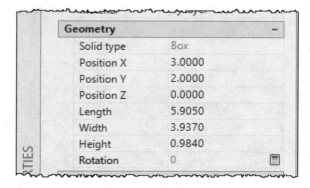

Inch dimensions after modifying.

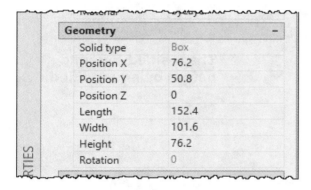

Metric dimensions before modifying.

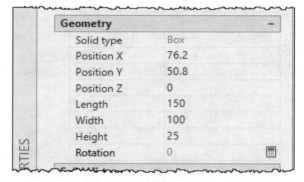

Metric dimensions after modifying.

8. After you have completed the changes press the **Escape** key on your keyboard to deselect the Box.

The 3D solid Box has now reduced in size to the dimensions you specified in the Properties Palette.

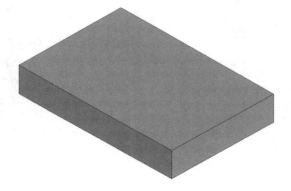

Modify Solids Using the Properties Palette....continued

As well as changing the sizes of 3D solids in the Properties Palette, you can also change the number of sides in a solid Pyramid. You can also change the bottom and top radius of a Pyramid or Cone or whether it is inscribed or circumscribed around a radius.

The example below shows a 6-sided Pyramid with an apex (sharp point) that has been modified to an 8-sided Pyramid with a top radius added to make it a truncated Pyramid.

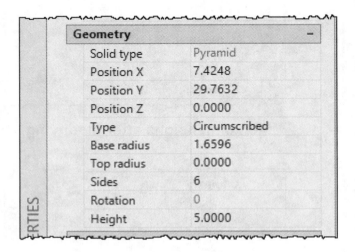

Geometry	−
Solid type	Pyramid
Position X	7.4248
Position Y	29.7632
Position Z	0.0000
Type	Circumscribed
Base radius	1.6596
Top radius	0.0000
Sides	6
Rotation	0
Height	5.0000

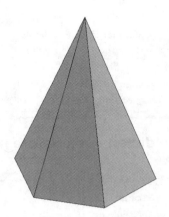

The 6-sided Pyramid before being modified.

Geometry	−
Solid type	Pyramid
Position X	7.4248
Position Y	29.7632
Position Z	0.0000
Type	Circumscribed
Base radius	1.7705
Top radius	0.4854
Sides	8
Rotation	0
Height	5.0000

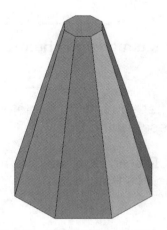

The 6-sided Pyramid has now been modified to an 8-sided Pyramid with a truncated top.

Modify Solids Using the Grips

Using the **Grips** on a 3D solid model is one of the quickest methods you can use to modify the models sizes. After selecting a 3D solid model like a Wedge, a serious of square and arrow Grips appear.

The square Grips allow you to change the overall length and width of the base individually or at the same time. The arrow Grips allow you to change the length, width and height individually.

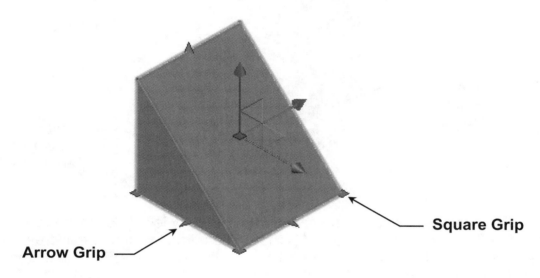

Square Grip

Arrow Grip

If you hover your mouse cursor over a square Grip it will show the current sizes of the length and width of the base. In the example below the mouse cursor is hovering over a square Grip showing the current length and width of the base in the "X" and "Y" axis.

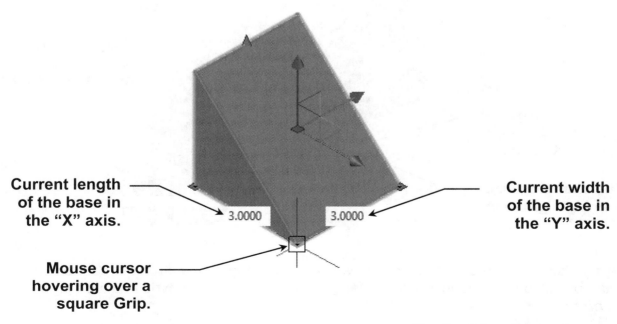

Current length of the base in the "X" axis.

3.0000   3.0000

Current width of the base in the "Y" axis.

Mouse cursor hovering over a square Grip.

Continued on the next page...

Modify Solids Using the Grips....continued

If you hover your mouse cursor over an arrow Grip it will show the current size in the direction of the arrow. In the example below the mouse cursor is hovering over an arrow Grip showing the length of the base in the "X" axis.

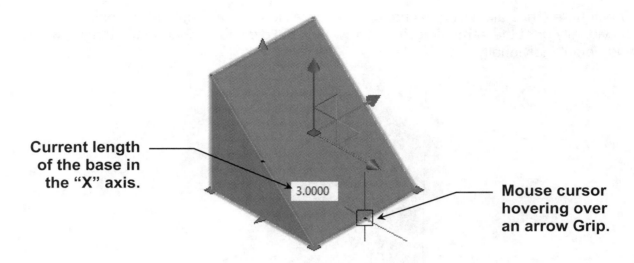

Current length of the base in the "X" axis.

3.0000

Mouse cursor hovering over an arrow Grip.

To change a particular size you must **activate** a Grip by left clicking with your mouse. The Grip will turn red, which means it is "hot". Then drag the cursor to increase or decrease the size. The changing size will be highlighted in blue. Type in the new size required then press **<enter>**. **Note:** It is best to have Ortho Mode turned on (**F8**).

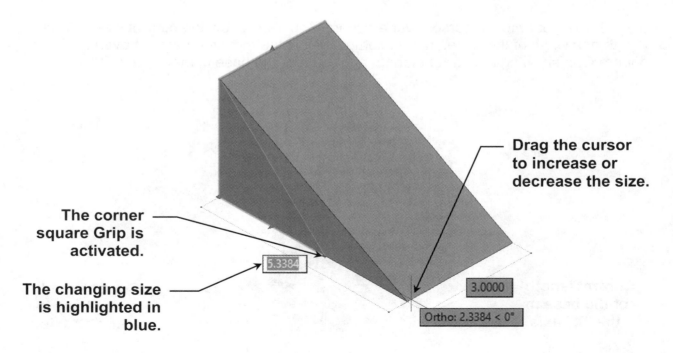

Drag the cursor to increase or decrease the size.

The corner square Grip is activated.

5.3384

The changing size is highlighted in blue.

3.0000

Ortho: 2.3384 < 0°

Note: Use the square Grips if you know the overall length or width required. For example, if you want to increase the length from 3" to 5" simply type in 5 then press **<enter>**.

Continued on the next page...

Modify Solids Using the Grips....continued

In the example below the arrow Grip has been activated to change the length of the Wedge. The amount of increase or decrease to the overall size is highlighted in blue.

Type in the amount you want to increase or decrease the overall length by then press **<enter>**. <u>Note</u>: It is best to have Ortho Mode turned on (**F8**).

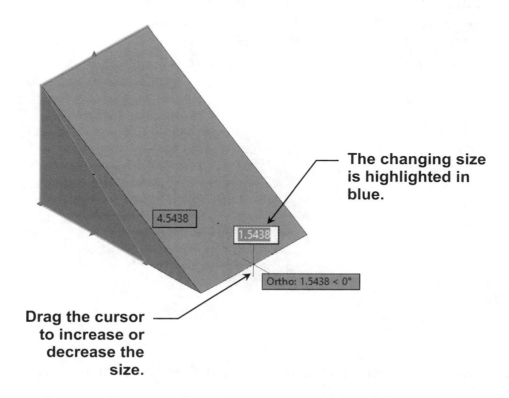

4.5438

1.5438

Ortho: 1.5438 < 0°

The changing size is highlighted in blue.

Drag the cursor to increase or decrease the size.

<u>Note</u>: Use the arrow Grips if you know the amount you want to increase or decrease the overall length, width or height by. For example, if you want to increase the length by 1.285" drag the cursor so it increases in length then simply type in 1.285 then press **<enter>**.

You can use the arrow Grips to change the height or radius of a Cylinder or to change the radius of a Sphere.

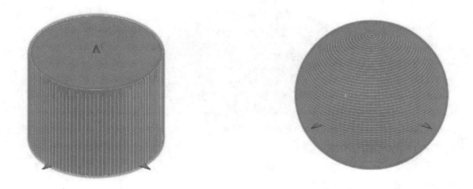

Exercise 2A

1. Start a new drawing file by selecting either **acad.dwt** for inch users, or **acadiso.dwt** for metric users.

2. Create a solid **Box** with the following sizes.

    Inch users

    Length = 4.375"
    Width = 3.850"
    Height = 2.900"

    Metric users

    Length = 111.12 mm
    Width = 97.79 mm
    Height = 73.66 mm

3. Using the **Properties Palette**, change the length, width and height to the following sizes.

    Inch users

    Length = 4.000"
    Width = 3.000"
    Height = 1.500"

    Metric users

    Length = 101.6 mm
    Width = 76.2 mm
    Height = 38.1 mm

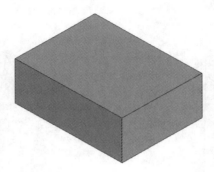

4. Save the drawing file as **Ex-2A**

Exercise 2B

1. Start a new drawing file by selecting either **acad.dwt** for inch users, or **acadiso.dwt** for metric users.

2. Create a solid **Pyramid** with the following sizes.

Inch users

Sides = 6
Circumscribed
Base radius = 1.575"
Height = 4.134"

Metric users

Sides = 6
Circumscribed
Base radius = 40 mm
Height = 105 mm

3. Using the **Properties Palette**, change the number of sides and add a top radius to the following sizes.

Inch users

Sides = 8
Top radius = 0.590"

Metric users

Sides = 8
Top radius = 15 mm

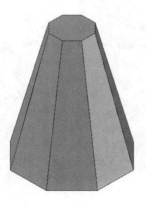

4. Save the drawing file as **Ex-2B**

Exercise 2C

1. Start a new drawing file by selecting either **acad.dwt** for inch users, or **acadiso.dwt** for metric users.

2. Create a solid **Wedge** with the following sizes.

    Inch users

    Length = 3.500"
    Width = 2.500"
    Height = 2.000"

    Metric users

    Length = 88.9 mm
    Width = 63.5 mm
    Height = 50.8 mm

3. Using the **Grips**, change the length, width and height to the following sizes.

    Inch users

    Length = 5.250"
    Width = 4.300"
    Height = 3.950"

    Metric users

    Length = 133.35 mm
    Width = 109.22 mm
    Height = 100.33 mm

4. Save the drawing file as **Ex-2C**

Exercise 2D

1. Start a new drawing file by selecting either **acad.dwt** for inch users, or **acadiso.dwt** for metric users.

2. Create a solid **Cylinder** with the following sizes.

   Inch users

   Base radius = 1.496"
   Top radius = 0.472"
   Height = 2.165"

   Metric users

   Base radius = 38 mm
   Top radius = 12 mm
   Height = 55 mm

3. Using the **Grips**, change the base radius, top radius and height to the following sizes.

   Inch users

   Base radius = 1.575"
   Top radius = 1.181"
   Height = 1.968"

   Metric users

   Base radius = 40 mm
   Top radius = 30 mm
   Height = 50 mm

Note: Remember to add on the amount to the existing sizes. For example, to increase the base radius from 1.496" [38 mm] to 1.575" [40 mm], move the grip to enlarge the base radius then type in 0.079" [2 mm] and then press **<enter>**.

4. Save the drawing file as **Ex-2D**

Exercise 2E

1. Start a new drawing file by selecting either **acad.dwt** for inch users, or **acadiso.dwt** for metric users.

2. Create a solid **Torus** with the following sizes.

Inch users

Torus radius = 1.500"
Tube radius = 0.500"

Metric users

Torus radius = 38.1 mm
Tube radius = 12.7 mm

3. Using either the **Properties Palette** or the **Grips**, change the Tube radius to the following size.

Inch users

Tube radius = 0.250"

Metric users

Tube radius = 6.35 mm

4. Save the drawing file as **Ex-2E**

LEARNING OBJECTIVES

After completing this lesson you will be able to:

1. Set the Solid History System Variable.
2. Chamfer the Edge of a 3D Solid Model.
3. Modify an Existing Chamfer.
4. Remove an Existing Chamfer.
5. Fillet the Edge of a 3D Solid Model.
6. Modify an Existing Fillet.
7. Remove an Existing Fillet.

LESSON 3

Setting the Solid History System Variable

The **Solid History System Variable** allows you to retain the history of the original 3D solid model. If the System Variable is turned **off** and you modify a solid, by adding a Chamfer or a Fillet for example, you will not be able to modify the sizes of the Chamfer or Fillet in the Properties Palette. So it is good practice to turn the System Variable **on**.

By default the **Solid History System Variable** is turned **off**. To turn it on use any of the following methods.

<u>Note:</u> When turned on, the Solid History System Variable **does not** work for existing 3D solid models, so it is also good practice to turn it **on** before you create any 3D solid models.

Method 1 – Modeling Panel on the 3D Tools Tab of the Drafting and Annotation Workspace.

1. Left click on the **Solid History** button. It will turn blue. (Blue is **on**.)

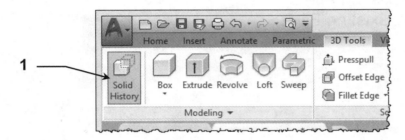

Method 2 – Primitive Panel on the Solid Tab of the 3D Modeling Workspace.

1. Left click on the **Solid History** button. It will turn blue. (Blue is **on**.)

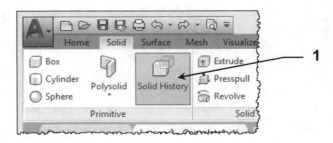

Method 3 – Keyboard entry.

1. On the **Command Line** or in the **Dynamic Input box**, type in *solidhist* on your keyboard then press *<enter>*.

2. Enter new value for SOLIDHIST <0>: *type in 1 then press <enter>*. (<u>Note:</u> **1** turns it **on** and **0** turns it **off**.)

Chamfer the Edge of a 3D Solid Model

The **Chamfer Edge** tool allows you to create a chamfer on the edges of 3D solid models. You can create an equal-sided chamfer or an unequal-sided chamfer on the outside or inside edges of 3D models, as shown in the examples below.

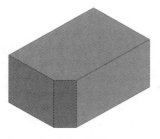

Equal-sided chamfer.

Unequal-sided chamfer.

Equal-sided chamfer.

Unequal-sided chamfer.

Further examples of chamfering 3D solid models.

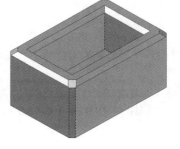

Chamfer the Edge of a 3D Solid Model....continued

There are various methods you can use to initiate the **Chamfer Edge** tool.

Method 1 – Solid Editing Panel on the 3D Tools Tab of the Drafting and Annotation Workspace.

1. Select the drop-down arrow on the **Fillet Edge** tool.

2. Select the **Chamfer Edge** tool from the list.

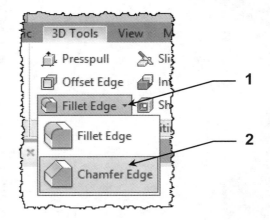

Method 2 – Solid Editing Panel on the Solid Tab of the 3D Modeling Workspace.

1. Select the drop-down arrow on the **Fillet Edge** tool.

2. Select the **Chamfer Edge** tool from the list.

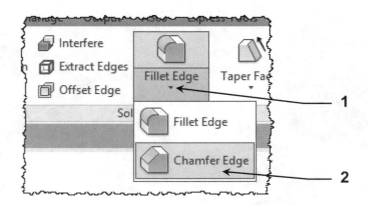

Method 3 – Keyboard entry.

1. On the **Command Line** or in the **Dynamic Input box**, type in *chamferedge* on your keyboard then press *<enter>*.

Chamfer the Edge of a 3D Solid Model....continued

An example of creating an equal chamfer on a single edge.

A 3D solid Box has been created with a length of 1.968" [50 mm], a width of 0.984" [25 mm], and a height of 0.984" [25 mm]. Create a chamfer on one end with equal sides of 0.236" x 0.236" [6 mm x 6 mm].

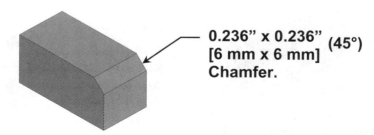

0.236" x 0.236" (45°)
[6 mm x 6 mm]
Chamfer.

1. Select the **Chamfer Edge** tool.
2. Select an edge or [Loop/Distance]: *left click on the top right-hand edge*.

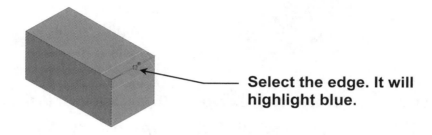

Select the edge. It will highlight blue.

3. Select an edge or [Loop/Distance]: *type in D* (for Distance) *then press <enter>*.
4. Specify Distance1 or [Expression] <1.000>: *type in 0.236 [6] then press <enter>*.
5. Specify Distance2 or [Expression] <1.000>: *type in 0.236 [6] then press <enter>*.
6. Select another edge on the same face or [Loop/Distance]: *press <enter>*.

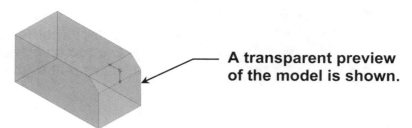

A transparent preview of the model is shown.

7. Press Enter to accept the chamfer or [Distance]: *press <enter> again*.

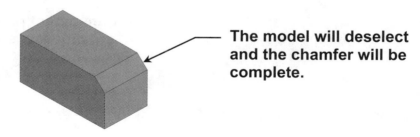

The model will deselect and the chamfer will be complete.

Chamfer the Edge of a 3D Solid Model….continued

An example of creating an equal chamfer on 4 edges of the same face.

A 3D solid Box has been created with a length of 1.968" [50 mm], a width of 0.984" [25 mm], and a height of 0.590" [15 mm]. Create a chamfer on the top 4 edges with equal sides of 0.118" x 0.118" [3 mm x 3 mm].

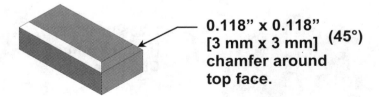

0.118" x 0.118"
[3 mm x 3 mm] (45°)
chamfer around
top face.

1. Select the **Chamfer Edge** tool.

2. Select an edge or [Loop/Distance]: *left click on the top right-hand edge.*

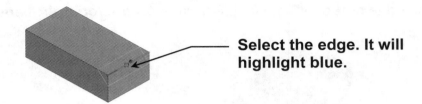

Select the edge. It will highlight blue.

3. Select an edge or [Loop/Distance]: *type in D* (for Distance) *then press <enter>.*

4. Specify Distance1 or [Expression] <0.236>: *type in 0.118 [3] then press <enter>.*

5. Specify Distance2 or [Expression] <0.236>: *type in 0.118 [3] then press <enter>.*

6. Select another edge on the same face or [Loop/Distance]: *type in L* (for Loop) *then press <enter>.*

7. Select edge of loop or [Edge/Distance]: *left click on another edge on the top face.*

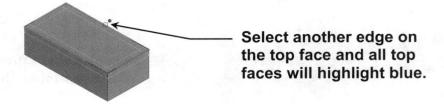

Select another edge on the top face and all top faces will highlight blue.

8. Select edge of loop or [Edge/Distance]: *press <enter>.*

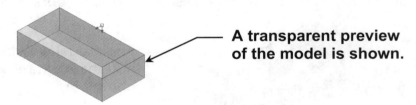

A transparent preview of the model is shown.

9. Press Enter to accept the chamfer or [Distance]: *press <enter> again to finish.*

Chamfer the Edge of a 3D Solid Model....continued

An example of creating an unequal chamfer on a single edge.

A 3D solid Cylinder has been created with a radius of 0.625" [15.87 mm], and a height of 1.500" [38.1 mm]. Create a chamfer on the top edge with unequal sides of 0.200" x 0.400" [5.08 mm x 10.16 mm].

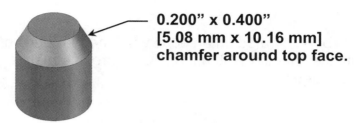

0.200" x 0.400"
[5.08 mm x 10.16 mm]
chamfer around top face.

1. Select the **Chamfer Edge** tool.

2. Select an edge or [Loop/Distance]: *left click on the top right-hand edge*.

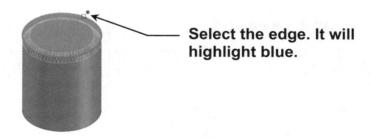

Select the edge. It will
highlight blue.

3. Select an edge or [Loop/Distance]: *type in D (for Distance) then press <enter>*.

4. Specify Distance1 or [Expression] <1.000>: *type in 0.200 [5.08] then press <enter>*.

5. Specify Distance2 or [Expression] <1.000>: *type in 0.400 [10.16] then press <enter>*.

6. Select another edge on the same face or [Loop/Distance]: *press <enter>*.

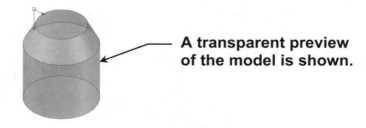

A transparent preview
of the model is shown.

7. Press Enter to accept the chamfer or [Distance]: *press <enter> again*.

Note: To have the unequal-sided chamfer the other way round, just reverse the sizes in **Steps 4-5**. Alternatively you can enter *D* at **Step 7** and change the sizes.

Modify an Existing Chamfer

You can modify an existing chamfer either by using the **Properties Palette** or by using the **Grips**. <u>Note</u>: **Solid History** must be turned on before you create the 3D solid model otherwise you will not be able to use the Properties Palette to modify the chamfer.

An example of modifying an existing chamfer using the Properties Palette.

A chamfer has been created on an edge with unequal sides of 0.118" x 0.315" [3 mm x 8 mm]. Modify the chamfer so that both sides are of equal length at 0.118" [3 mm] using the Properties Palette.

1. Open the **Properties Palette**.

2. Zoom into the chamfer with the wheel on your mouse then hold down the *Ctrl* key on your keyboard and left click on the chamfer. The chamfer will highlight blue.

3. Left click on the Second Distance box in the **Properties Palette** and change the size from 0.315 [8] to 0.118 [3] then press **<enter>**.

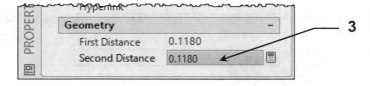

4. Press the *Escape* key on your keyboard to deselect the chamfer.

The chamfer now has equal sides of 0.118" [3 mm].

Modify an Existing Chamfer....continued

An example of modifying an existing chamfer using the Grips.

A chamfer has been created on an edge with equal sides of 0.118" x 0.118" [3 mm x 3 mm]. Modify the chamfer so that the top face distance is 0.315" [8 mm] using the Grips.

1. Zoom into the chamfer with the wheel on your mouse then hold down the **Ctrl** key on your keyboard and left click on the chamfer. The chamfer will highlight blue.

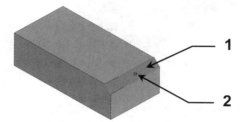

2. Left click on the blue down arrow. The 2 chamfer distance grips will appear.

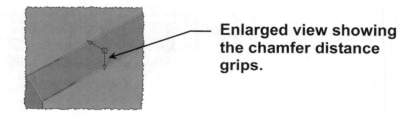

Enlarged view showing the chamfer distance grips.

3. Left click on the top arrow to select it, then drag it to the left. Type in 0.315 [8] and then press **<enter>**.

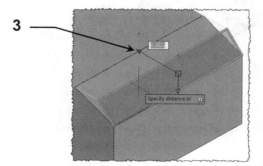

4. Press the **Escape** key on your keyboard to deselect the chamfer.

The chamfer now has a top face distance of 0.315" [8 mm], with the end face distance remaining at 0.118" [3 mm].

Remove an Existing Chamfer

If you want to completely remove an existing chamfer from a 3D solid model there are 2 methods you can use. You either can change the first and second distances to zero in the **Properties Palette**, or you can use the *Delete* key on your keyboard.

Remove an existing chamfer using the Properties Palette.

1. Open the **Properties Palette**.

2. Zoom into the chamfer with the wheel on your mouse then hold down the *Ctrl* key on your keyboard and left click on the chamfer. The chamfer will highlight blue.

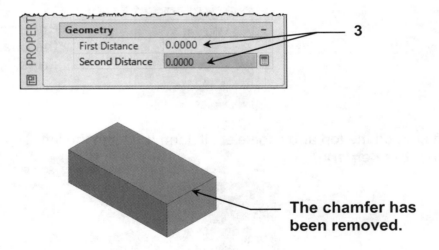

3. Left click on the First Distance box in the **Properties Palette** and change the size to 0 (zero). Left click in the Second Distance box and change the size to 0 (zero) then press *<enter>*. The chamfer has now been removed.

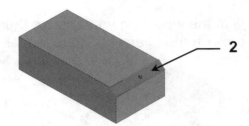

The chamfer has been removed.

Remove an existing chamfer using the Delete key.

1. Zoom into the chamfer with the wheel on your mouse then hold down the *Ctrl* key on your keyboard and left click on the chamfer. The chamfer will highlight blue.

2. Press the *Delete* key on your keyboard. The chamfer has now been removed.

Fillet the Edge of a 3D Solid Model

The **Fillet Edge** tool allows you to create a rounded edge on 3D solid models. You can create a fillet on the outside or inside edges of 3D models, as shown in the examples below.

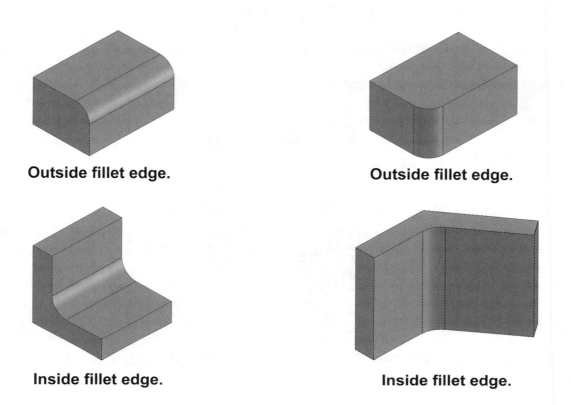

Outside fillet edge.   **Outside fillet edge.**

Inside fillet edge.   **Inside fillet edge.**

Further examples of filleting 3D solid models.

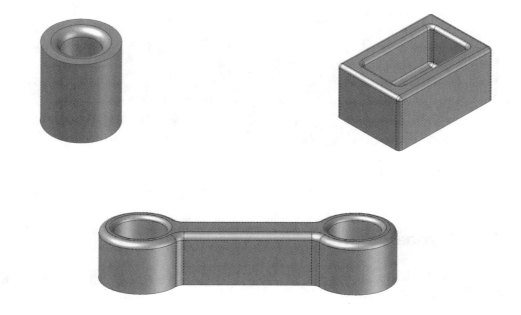

Fillet the Edge of a 3D Solid Model....continued

There are various methods you can use to initiate the **Fillet Edge** tool.

Method 1 – Solid Editing Panel on the 3D Tools Tab of the Drafting and Annotation Workspace.

1. Select the drop-down arrow on the **Fillet Edge** tool.

2. Select the **Fillet Edge** tool from the list.

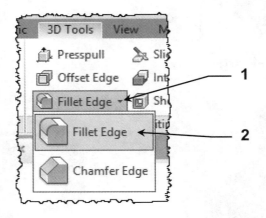

Method 2 – Solid Editing Panel on the Solid Tab of the 3D Modeling Workspace.

1. Select the drop-down arrow on the **Fillet Edge** tool.

2. Select the **Fillet Edge** tool from the list.

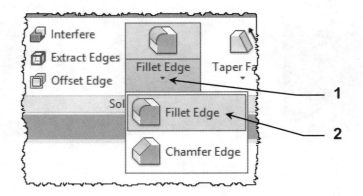

Method 3 – Keyboard entry.

1. On the **Command Line** or in the **Dynamic Input box**, type in *filletedge* on your keyboard then press *<enter>*.

Fillet the Edge of a 3D Solid Model....continued

An example of creating a fillet on a single edge.

A 3D solid Box has been created with a length of 1.000" [25.4 mm], a width of 1.500" [38.1 mm], and a height of 0.750" [19.05 mm]. Create a fillet on one end with a radius of 0.250" [6.35 mm].

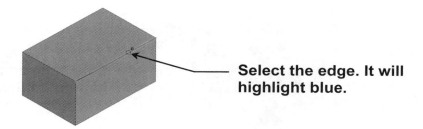

R0.250" [R6.35 mm] fillet.

1. Select the **Fillet Edge** tool.
2. Select an edge or [Chain/Loop/Radius]: *left click on the top right-hand edge*.

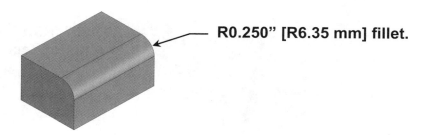

Select the edge. It will highlight blue.

3. Select an edge or [Chain/Loop/Radius]: *type in R* (for Radius) *then press <enter>*.
4. Enter fillet radius or [Expression] <0.050>: *type in 0.250 [6.35] then press <enter>*
5. Select an edge or [Chain/Loop/Radius]: *press <enter>*.

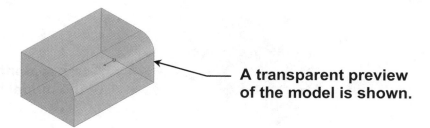

A transparent preview of the model is shown.

6. Press Enter to accept the fillet or [Radius]: *press <enter> again*.

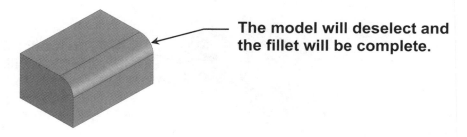

The model will deselect and the fillet will be complete.

Fillet the Edge of a 3D Solid Model....continued

An example of creating a fillet on 4 edges of the same face.

A 3D solid Box has been created with a length of 1.000" [25.4 mm], a width of 1.500" [38.1 mm], and a height of 0.750" [19.05 mm]. Create a fillet on the top 4 edges with a radius of 0.125" [3.17 mm].

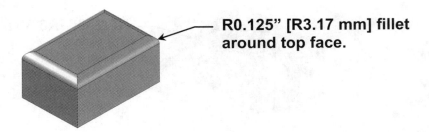

R0.125" [R3.17 mm] fillet around top face.

1. Select the **Fillet Edge** tool.

2. Select an edge or [Chain/Loop/Radius]: *left click on the top right-hand edge*.

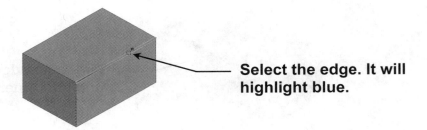

Select the edge. It will highlight blue.

3. Select an edge or [Chain/Loop/Radius]: *type in R* (for Radius) *then press <enter>*.

4. Enter fillet radius or [Expression] <0.050>: *type in 0.125 [3.17] then press <enter>*

5. Select an edge or [Chain/Loop/Radius]: *select the other 3 edges on the top face then press <enter>*.

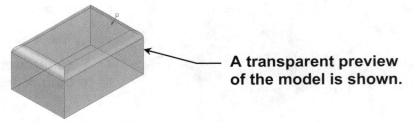

A transparent preview of the model is shown.

6. Press Enter to accept the fillet or [Radius]: *press <enter> again*.

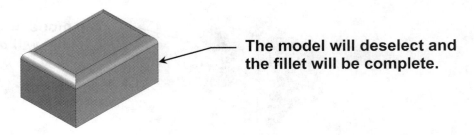

The model will deselect and the fillet will be complete.

Modify an Existing Fillet

You can modify an existing fillet either by using the **Properties Palette** or by using the **Grips**. **Note: Solid History** must be turned on before you create the 3D solid model otherwise you will not be able to use the Properties Palette to modify the fillet.

An example of modifying an existing fillet using the Properties Palette.

A fillet has been created on an edge with a radius of 0.250" [6.35 mm]. Modify the fillet by increasing the radius to 0.500" [12.7 mm] using the Properties Palette.

1. Open the **Properties Palette**.

2. Zoom into the fillet with the wheel on your mouse then hold down the **Ctrl** key on your keyboard and left click on the fillet. The fillet will highlight blue.

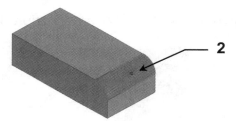

3. Left click on the Radius box in the **Properties Palette** and change the size from 0.250 [6.35] to 0.500 [12.7] then press **<enter>**.

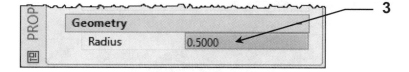

4. Press the **Escape** key on your keyboard to deselect the fillet.

The fillet has now increased in size to 0.500" [12.7 mm].

Modify an Existing Fillet....continued

An example of modifying an existing fillet using the Grips.

A fillet has been created on an edge with a radius of 0.250" [6.35 mm]. Modify the fillet by increasing the radius to 0.500" [12.7 mm] using the Grips.

1. Zoom into the fillet with the wheel on your mouse then hold down the **Ctrl** key on your keyboard and left click on the fillet. The fillet will highlight blue.

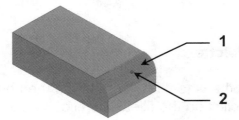

2. Left click on the blue down arrow. The fillet sizing grip will appear.

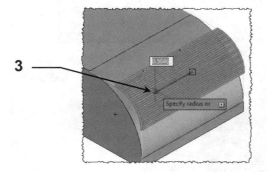

Enlarged view showing the fillet sizing grip.

3. Left click on the diagonal arrow to select it, then drag it down and to the left. Type in 0.500 [12.7] and then press **<enter>**.

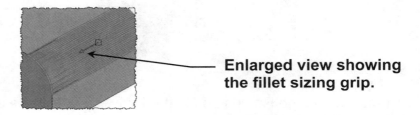

4. Press the **Escape** key on your keyboard to deselect the fillet.

The fillet has now increased in size to 0.500" [12.7 mm].

Remove an Existing Fillet

If you want to completely remove an existing fillet from a 3D solid model there are 2 methods you can use. You either can change the radius size to zero in the **Properties Palette**, or you can use the *Delete* key on your keyboard.

Remove an existing fillet using the Properties Palette.

1. Open the **Properties Palette**.

2. Zoom into the fillet with the wheel on your mouse then hold down the *Ctrl* key on your keyboard and left click on the fillet. The fillet will highlight blue.

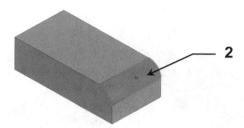

3. Left click on the Radius box in the **Properties Palette** and change the size to 0 (zero) then press *<enter>*. The fillet has now been removed.

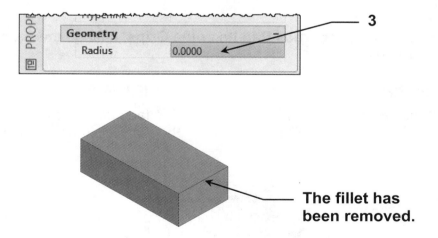

The fillet has been removed.

Remove an existing fillet using the Delete key.

1. Zoom into the fillet with the wheel on your mouse then hold down the *Ctrl* key on your keyboard and left click on the fillet. The fillet will highlight blue.

2. Press the *Delete* key on your keyboard. The fillet has now been removed.

Exercise 3A

1. Start a new drawing file by selecting either **acad.dwt** for inch users, or **acadiso.dwt** for metric users.

2. Create a solid **Box** with the following sizes, then chamfer the top right-hand edge.

<u>**Inch users**</u>

    Length = 3.937"
    Width   = 2.953"
    Height = 1.181"

    Chamfer size = 0.236" x 0.236"

<u>**Metric users**</u>

    Length = 100 mm
    Width   = 75 mm
    Height = 30 mm

    Chamfer size = 6 mm x 6 mm

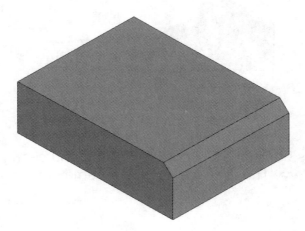

3. Using either the **Properties Palette** or the **Grips**, change the chamfer to the following sizes.

<u>**Inch users**</u>

    Top face distance = 0.787"
    Side face distance = 0.472"

<u>**Metric users**</u>

    Top face distance = 20 mm
    Side face distance = 12 mm

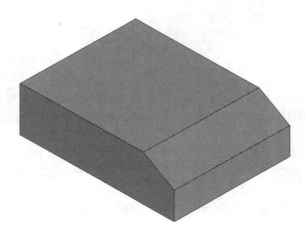

4. Save the drawing file as **Ex-3A**

Exercise 3B

1. Open the drawing file **Ex-3A**.

2. Remove the existing chamfer on the top right-hand edge using either the **Properties Palette** or the *Delete* key on your keyboard.

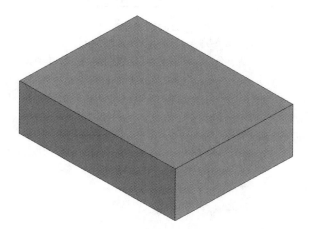

3. Chamfer the 4 edges on the top face to the following sizes.

Inch users

Top face distance = 0.200"
Side face distance = 0.400"

Metric users

Top face distance = 5.08 mm
Side face distance = 10.16 mm

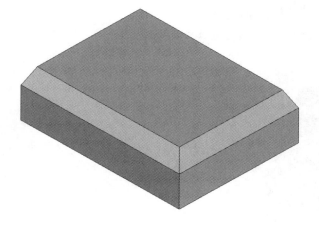

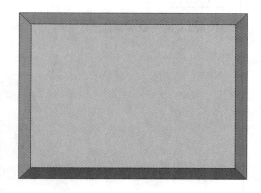

Top view showing the 4 chamfered edges on the top face.

4. Save the drawing file as **Ex-3B**

Exercise 3C

1. Start a new drawing file by selecting either **acad.dwt** for inch users, or **acadiso.dwt** for metric users.

2. Create a solid **Pyramid** with the following sizes, then fillet the top edges.

Inch users

Sides = 6
Circumscribed
Base radius = 1.000"
Top radius = 0.500"
Height = 2.000"

Fillet radius = 0.100"

Metric users

Sides = 6
Circumscribed
Base radius = 25.4 mm
Top radius = 12.7 mm
Height = 50.8 mm

Fillet radius = 2.54 mm

3. Using either the **Properties Palette** or the **Grips**, increase the fillets to the following sizes.

Inch users

Fillet radius = 0.250"

Metric users

Fillet radius = 6.35 mm

4. Save the drawing file as **Ex-3C**

LEARNING OBJECTIVES

After completing this lesson you will be able to:

1. Understand the User Coordinate System (UCS).
2. Move the UCS to a Temporary Location.
3. Rotate the UCS.
4. Use the Dynamic UCS.

LESSON 4

User Coordinate System (UCS)

In order for you to be able to construct more complex 3D models, it is essential that you learn how to move and manipulate the **User Coordinate System** or **UCS** for short.

World Coordinate System (WCS).

All the objects or 3D models you create are defined by the X,Y,Z coordinates that are measured from 0,0,0, or the **Origin**. This is a **fixed** position and referred to as the **World Coordinate System** or **WCS** for short.

User Coordinate System (UCS).

The **UCS** allows you to move a temporary **Origin** to any position you wish. This could be the corner of a 3D solid box or the center of a Cylinder. This is referred to as "Moving the Origin".

An example of why you would move the UCS to a temporary Origin.

A solid Wedge has been created with the base of the Wedge on the X,Y plane. The UCS icon is showing the length of the Wedge in the X axis, the width in the Y axis, and the height of the Wedge in the Z axis.

A solid Cylinder is to be placed on the sloped face and perpendicular (right angle) to that face. The UCS icon has been moved to the top left-hand corner with the X axis and Y axis now on the same plane as the sloped face.

The UCS icon was moved to the new location using the **UCS 3 Point** Command. You will learn the UCS 3 Point command in this lesson.

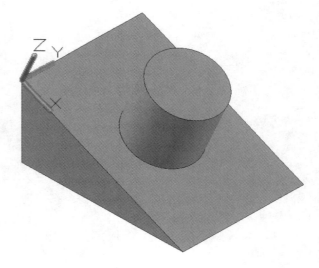

User Coordinate System (UCS)....continued

The UCS icon was moved to the top left-hand corner of the wedge by using the **UCS 3 Point** command and snapping to the corner using the **Endpoint Object Snap**. Object Snaps are used to snap to objects at specific and accurate locations.

To check you have the **Endpoint Snap** enabled, do the following.

1. Type in **DS** (lowercase is OK) on your keyboard and then press **<enter>**.

2. The **Drafting Settings** dialog box will open showing the Object Snap Tab.

3. Place a checkmark in the **Endpoint**, **Midpoint** and **Center** boxes. These are the 3 Object Snaps most commonly used throughout this book.

Note: AutoCAD version 2015 or earlier will not have the Geometric Center Snap, but that is OK as it will not be required in this book.

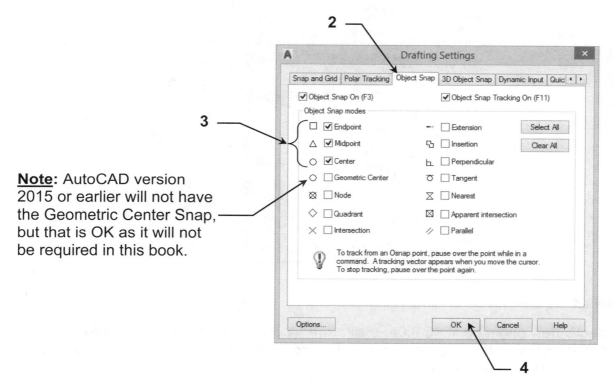

4. Select **OK** to close the Drafting Settings dialog box.

You can turn the Objects Snaps on or off by selecting the Object Snap icon on the **Status Bar** at the bottom of your screen, or by toggling the **F3** key on your keyboard. Blue is **ON** and Grey is **OFF**.

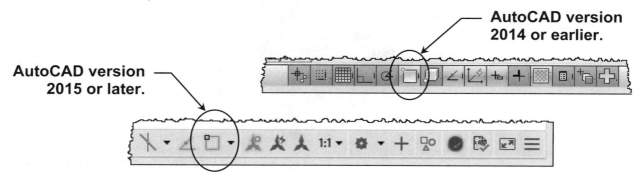

AutoCAD version 2014 or earlier.

AutoCAD version 2015 or later.

Moving the UCS to a Temporary Location

Moving the UCS using the 3 Point command.

In this example the UCS Origin needs to be moved on to the sloped face of a Wedge with the X axis and Y axis on the same plane as the sloped face.

1. Left click on the **UCS** icon to activate it.

2. Right click (press the right mouse button). The shortcut Menu shown below will appear.

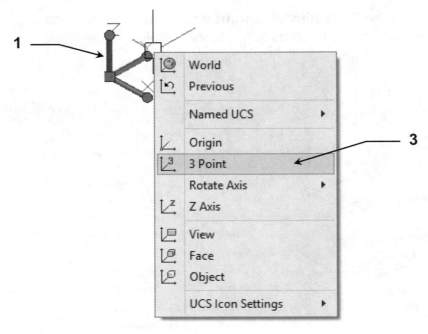

3. Select the **3 Point** command.

4. Specify new origin point <0,0,0>: *move the UCS icon to the top left-hand corner of the wedge until the green square Endpoint Snap appears, then left click to snap the UCS icon to that corner.*

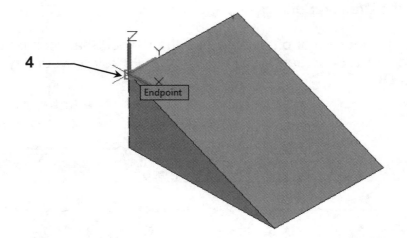

Continued on the next page...

Moving the UCS to a Temporary Location....continued

5. Specify point on positive portion of X axis <1.000,0.000,1.800>: *move your mouse to the bottom left-hand corner of the sloped face until the green square Endpoint Snap appears, then left click to snap the UCS icon to that corner.*

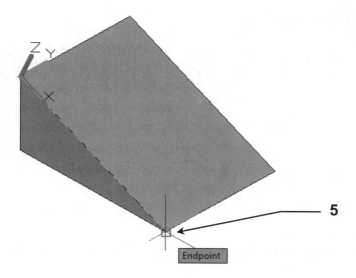

6. Specify point on positive Y portion of the UCS XY plane <0.000,1.000,1.800>: *move your mouse to the top right-hand corner of the sloped face until the green square Endpoint Snap appears, then left click to snap the UCS icon to that corner.*

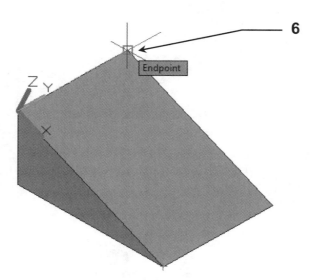

7. The **3 Point** command will end automatically.

The **UCS Origin** has now moved to the top left-hand corner of the Wedge with the X axis and Y axis on the same plane as the sloped face. The Z axis is now perpendicular to the sloped face.

Moving the UCS to a Temporary Location....continued

There are other methods you can use to move the UCS Origin. You can choose to just move the Origin only, or you can choose to move the Origin and align it with objects automatically.

To move the Origin only.

1. Left click on the **UCS** icon. The icon will highlight with 3 circle Grips on the positive ends of the each axis, and a single square Grip at the base of the UCS icon.

2. Hover your mouse over the square base Grip. A shortcut Menu will appear.

3. Select **Move Origin Only** from the Menu.

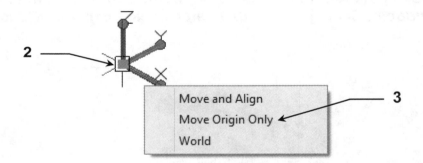

4. Move the UCS icon and snap the icon to any location on an object. For example, the center at one end of a Cylinder.

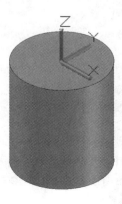

4-6

Moving the UCS to a Temporary Location....continued

To move and align the Origin.

1. Left click on the **UCS** icon. The icon will highlight with 3 circle Grips on the positive ends of the each axis, and a single square Grip at the base of the UCS icon.

2. Hover your mouse over the square base Grip. A shortcut Menu will appear.

3. Select **Move and Align** from the Menu.

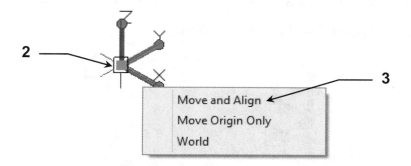

4. Move the UCS icon on to any surface of the 3D model. The surface will highlight blue.

5. When the UCS icon appears on the surface you require, left click your mouse. The surface will revert back to its original color.

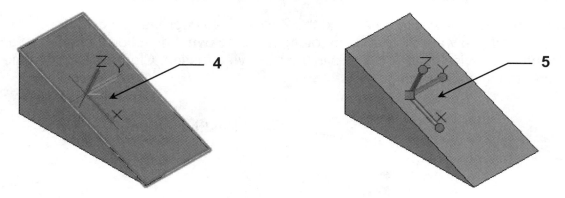

6. Press the *Escape* key on your keyboard to deselect the UCS icon.

To return to the **World Coordinate System (WCS)** at 0,0,0, select **World** from the Menu.

Rotating the UCS

The UCS can also be rotated around the X, Y, and Z axis. You may use your cursor to define the rotation angle or you can enter the rotation angle on your keyboard

An example of rotating the X axis around the Z axis.

1. Left click on the **UCS** icon. The icon will highlight with 3 circle Grips on the positive ends of the each axis, and a single square Grip at the base of the UCS icon.

2. Hover your mouse over the **X axis circle Grip**. A shortcut Menu will appear.

3. Select **Rotate Around Z Axis** from the Menu.

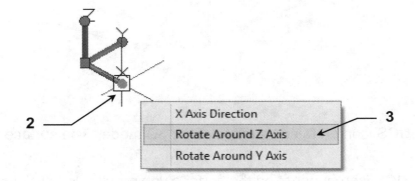

4. Specify rotation angle about Z axis <90>: *type in 90 then press <enter>*.

5. The X axis has now rotated around the Z axis by 90 degrees in a **counterclockwise** direction. If you want the X axis to rotate in a **clockwise** direction just enter **−90**. Counterclockwise rotates the axis in the positive direction. Clockwise rotates the axis in the negative direction.

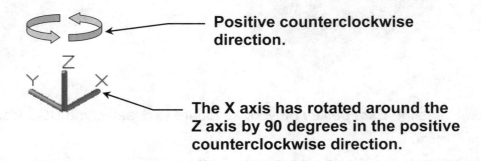

Positive counterclockwise direction.

The X axis has rotated around the Z axis by 90 degrees in the positive counterclockwise direction.

Note: You may specify any angle from **0** to **360** degrees. For example, **38.5**

Rotating the UCS....continued

An example of rotating the Y axis around the X axis.

1. Left click on the **UCS** icon. The icon will highlight with 3 circle Grips on the positive ends of the each axis, and a single square Grip at the base of the UCS icon.

2. Hover your mouse over the **Y axis circle Grip**. A shortcut Menu will appear.

3. Select **Rotate Around X Axis** from the Menu.

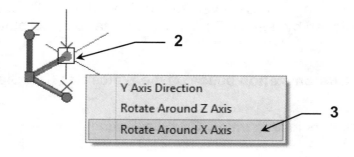

4. Specify rotation angle about X axis <90>: *type in 90 then press <enter>*.

5. The Y axis has now rotated around the X axis by 90 degrees in a **counterclockwise** direction. If you want the Y axis to rotate in a **clockwise** direction just enter **−90**. Counterclockwise rotates the axis in the positive direction, clockwise rotates the axis in the negative direction.

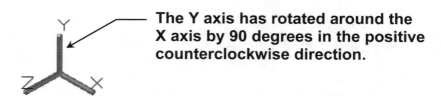

The Y axis has rotated around the X axis by 90 degrees in the positive counterclockwise direction.

Note: You can rotate any axis around any other axis using this method.

To return to the **World Coordinate System** (**WCS**) at 0,0,0, type in *ucs* on your keyboard then press *<enter>,* then press *<enter>* again.

Using the Dynamic UCS

The **Dynamic UCS** allows you temporarily align the X,Y plane of the UCS to a face on a 3D solid object. The Dynamic UCS becomes active **after** you have initiated a command, such as the Cylinder command. You can turn the Dynamic UCS on or off by toggling the **F6** key on your keyboard, or by selecting the **Dynamic UCS** button on the Status Bar.

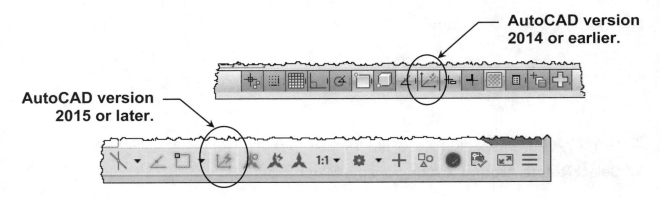

AutoCAD version 2014 or earlier.

AutoCAD version 2015 or later.

Note: If you are using AutoCAD version 2015 or later the **Dynamic UCS** button on the Status Bar is hidden by default. To show the Dynamic UCS button on the Status Bar, do the following.

1. Select the **Customization** button on the Status Bar. The Status Bar Menu will appear.

2. Put a checkmark in **Dynamic UCS** by selecting it.

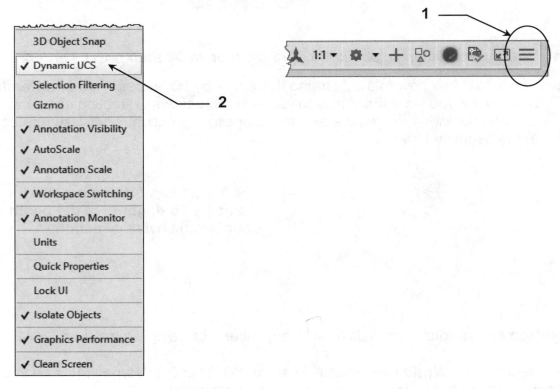

3. Left click in the drawing area to close the Status Bar Menu.

Using the Dynamic UCS....continued

An example of using the Dynamic UCS.

A solid Cylinder needs to be placed on the end of a solid box. The Cylinder can be placed without having to rotate or align the UCS.

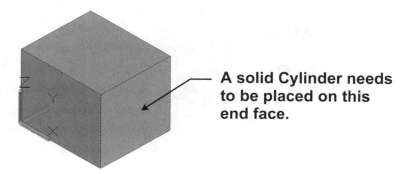

A solid Cylinder needs to be placed on this end face.

1. Select the **Cylinder** tool.

2. Move the mouse cursor on to the right-hand end face in the approximate position you would like the center of the Cylinder to be placed.

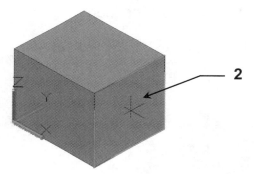

2

3. Continue with the **Cylinder** command by using a left click to place the Cylinder, then entering the radius and height of the Cylinder.

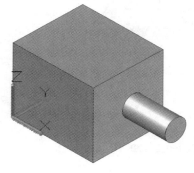

Note: The Cylinder was placed in an approximate position and you may wish to move it to a more accurate position on the end face. Moving 3D solids will be shown in later lessons.

Exercise 4A

1. Open **Ex-1A**.

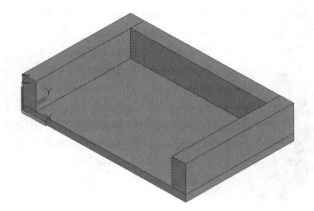

2. Left click on the UCS icon to select it.

3. Hover your mouse over the **X axis circle Grip**. A shortcut Menu will appear.

4. Select **Rotate Around Z Axis** from the Menu.

5. Specify rotation angle about Z axis <90>: *type in –90 and then press <enter>*.

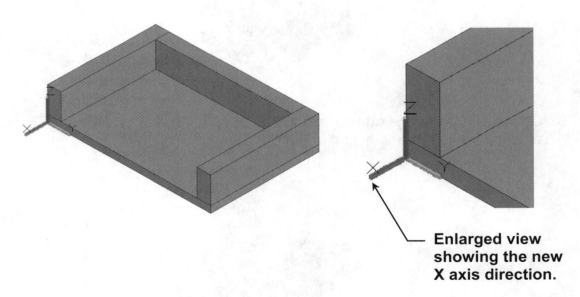

Enlarged view
showing the new
X axis direction.

6. Press the *Escape* key to deselect the UCS icon.

7. Save the drawing file as **Ex-4A** but keep the drawing file open.

Continued on the next page...

Exercise 4A....continued

A solid Wedge is to be placed on top of the 4 boxes, with the base size of the Wedge the same as the base size of the bottom of the 4 boxes, and with the slope of the Wedge running in the same direction as the positive X axis.

1. Select the **Wedge** tool.

2. Specify first corner or [Center]: *left click on the top left-hand corner using the Endpoint Object Snap at P1*.

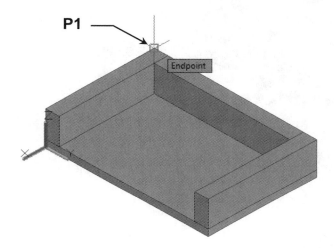

3. Specify other corner or [Cube/Length]: *drag your mouse down and to the right then left click on the diagonal corner using the Endpoint Object Snap at P2*.

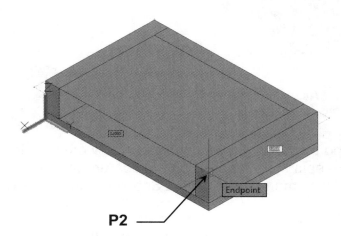

Continued on the next page...

Exercise 4A....continued

4. Specify height or [**2P**oint]: *drag your mouse up and to the right at P3, then type in 0.5 [12.7] for the height of the Wedge, and then press <enter>.*

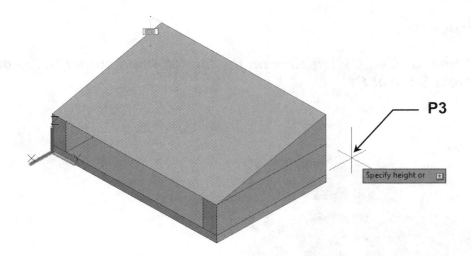

P3

Specify height or

5. The command will end automatically and the Wedge will be complete.

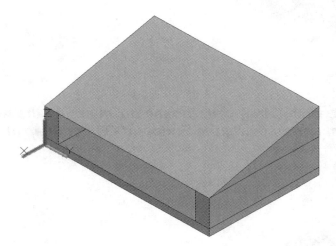

6. Resave the drawing file as **Ex-4A**.

You will have noticed that by using the Endpoint Object Snap, you were able to create the base size of the Wedge by snapping to endpoints on the corners of other 3D objects. You can use this method with many other Object Snaps including the **Center** and **Midpoint** Snaps.

The UCS was rotated around the Z axis by –90 degrees to enable the slope of the Wedge to run in the same direction as the positive X axis.

Exercise 4B

1. Open **Ex-4A** if it is not already open.

2. Immediately save the drawing file as **Ex-4B**.

A solid Cylinder is to be placed on the slope of the Wedge at the midpoint of the slope using the **Dynamic UCS**. The easiest way to accurately place the Cylinder at the midpoint of the Wedge is to create a guide line from one diagonal corner to the other. The center of the Cylinder can then be snapped to the midpoint of the guide line using the **Midpoint Snap**.

How to create the guide line.

1. Type in **L** on your keyboard (for the **Line** command) and then press **<enter>**.

2. Specify first point: *using the Endpoint Snap, left click on the bottom left-hand corner of the Wedge at P1*.

3. Specify next point or [Undo]: *using the Endpoint Snap, left click on the top right-hand corner of the Wedge at P2*.

4. Specify next point or [Undo]: *press <enter> to finish the Line command*.

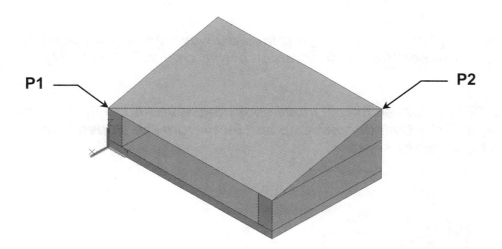

5. Resave the drawing file as **Ex-4B** but keep the drawing file open.

Continued on the next page...

Exercise 4B....continued

Creating the solid Cylinder using the Dynamic UCS.

Before creating the Cylinder make sure you have the **Midpoint** and **Center** Snaps enabled and the Object Snaps are turned on (refer to page 4-3). Also make sure you have the **Dynamic UCS** turned on (refer to page 4-10).

1. Select the **Cylinder** tool.

2. Specify center point of base or [3P/2P/Ttr/Elliptical]: *move your mouse to the center of the guide line until you see the triangle shaped Midpoint Snap, then left click to place the center of the Cylinder*.

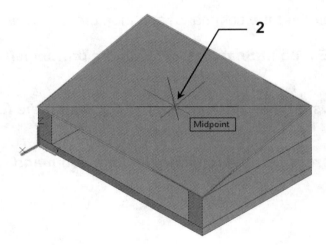

3. Specify base radius or [Diameter]: *move your mouse away from the midpoint of the guide line until you see the Cylinder radius begin, type in 0.625 [15.87] then press <enter>*.

4. Specify height or [2Point/Axis endpoint]: *move your mouse up and to the left until you see the height of the Cylinder move up and away from the sloped face, type in 0.5 [12.7] then press <enter>*.

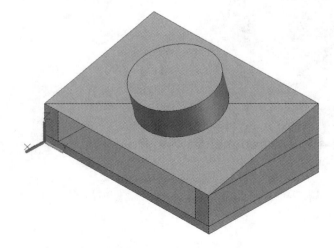

Continued on the next page...

Exercise 4B....continued

You may wish to remove the guide line that was used to place the center of the Cylinder at the midpoint of the sloped face of the Wedge. It is sometimes difficult to select the line when it is placed on a 3D solid model. The easiest way to select the line is to use a **Window Selection** to highlight just the line.

How to select the line using the Window Selection method.

1. Single left click with your mouse approximately at **P1**. (Make sure you release the mouse button.)

2. Move your mouse over to approximately **P2** then single left click with your mouse.

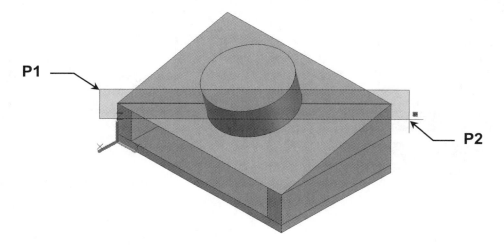

As you move your mouse over to **P2** you will see a blue color rectangle known as the Selection Window. The guide line will highlight. Only the objects enclosed in the Selection Window will be selected. So you have to be careful not to make the window too big in case the Cylinder is selected as well.

3. The Selection Window will disappear and the guide line will highlight blue.

4. Press the **Delete** key on your keyboard. The line will be removed.

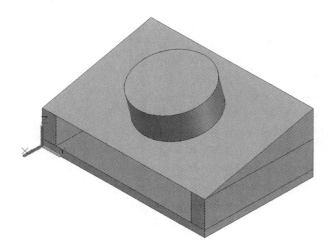

5. Resave the drawing file as **Ex-4B**.

Notes:

LEARNING OBJECTIVES

After completing this lesson you will be able to:

1. View your Drawing Using the Orbit Tool.
2. Move 3D Models Using Object Snaps.
3. Combine 3D Models Using the Union Tool.
4. Create 3D Models Using the Subtract Tool.
5. Create 3D Models Using the Intersect Tool.

LESSON 5

Orbit Tool

The **Orbit** tool allows you to rotate around a 3D solid model using the mouse click and drag method. There are 3 methods you can use with the Orbit tool, which are explained below.

Orbit

Constrains 3D Orbit along the **XY** plane or the **Z** axis. Click and hold down the left mouse button then drag the cursor to rotate the model.

Free Orbit

Orbits in any direction without reference to the planes. The point of view is not constrained along the **XY** plane of the **Z** axis. Click and hold down the left mouse button then drag the cursor to rotate the model.

Continuous Orbit

Orbits continuously. Click and drag in the direction you want the continuous orbit to move, and then release the mouse button. The orbit continues to move in that direction.

There are various methods you can use to initiate the **Orbit** tool.

Method 1 – Navigate Panel on the View Tab of the Drafting and Annotation Workspace.

Note: The **Navigate** Panel is off by default. Select the **View** Tab then right click on any Panel and select **Show Panels**. Then select **Navigate** from the list.

1. Select the **Orbit** drop-down arrow.

2. Select one of the 3 options from the list.

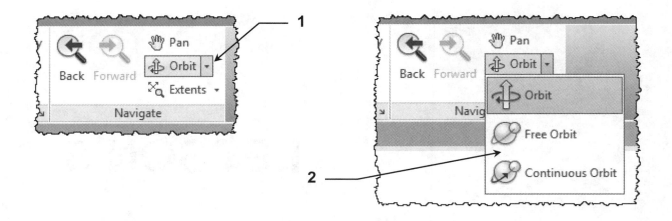

5-2

Orbit Tool....continued

Method 2 – Navigate Panel on the View Tab of the 3D Modeling Workspace.

Note: The **Navigate** Panel is off by default. Select the **View** Tab then right click on any Panel and select **Show Panels**. Then select **Navigate** from the list.

1. Select the **Orbit** drop-down arrow.

2. Select one of the 3 options from the list.

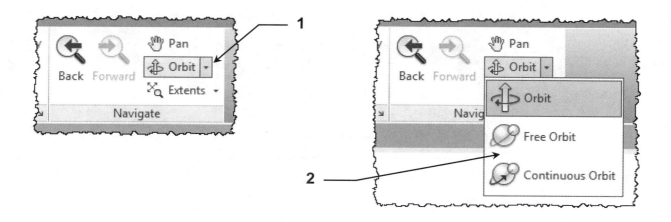

Method 3 – Navigation Bar in the Main Drawing Area.

1. Select the **Orbit** drop-down arrow.

2. Select one of the 3 options from the list.

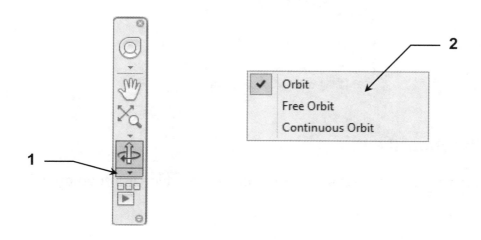

To end the **Orbit** command press either the *Esc* key or press *<enter>*.

Move Tool

The **Move** tool allows you to move a 3D solid model from its current location to a new location by using a base point and a displacement point. The Object Snaps you used in Lesson 4 are very useful when moving 3D solid models.

Selecting the Move tool.

Method 1 – Modify Panel on the Home Tab of the Drafting and Annotation Workspace.

1. Select the **Move** tool.

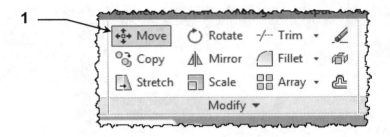

Method 2 – Modify Panel on the Home Tab of the 3D Modeling Workspace.

1. Select the **Move** tool.

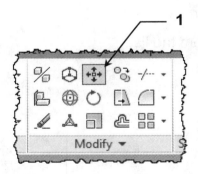

Method 3 – Keyboard entry.

1. On the **Command Line** or in the **Dynamic Input box**, type in *move* on your keyboard then press *<enter>*.

Move Tool....continued

An example of using the Move tool.

The midpoint of the small Box needs to be placed on the midpoint of the larger Box.

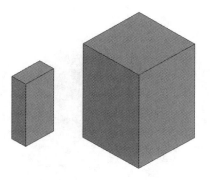 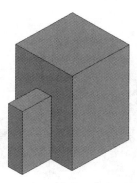

Note: You may find it easier to switch to **2D Wireframe** view when moving 3D solid models. Turning off **Ortho Mode** (*F8*) makes it easier to move objects.

1. Select the **Move** tool.

2. Select objects: *left click on the small Box and then press <enter>*.

3. Specify base point or [**D**isplacement] <Displacement>: *using the Midpoint Object Snap, left click on the midpoint of the bottom back edge of the small Box (P1).*

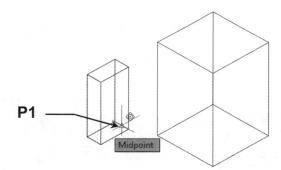

4. Specify second point or <use first point as displacement>: *drag your mouse cursor over to the front bottom edge of the large Box and using the Midpoint Object Snap, left click on the midpoint (P2).*

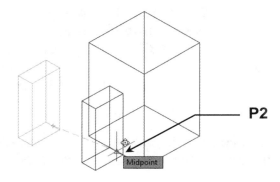 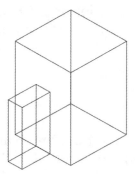

5. The **Move** command will end automatically.

Union Tool

The **Union** tool allows you to combine 2 or more 3D solid models to create one 3D solid model. Some examples of combining 3D solid models are shown below.

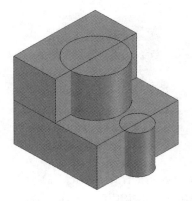

2 Cylinders and 2 Boxes need to be combined.

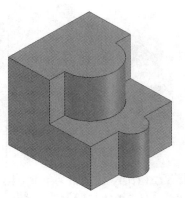

The 4 objects have now been combined into a single 3D solid model.

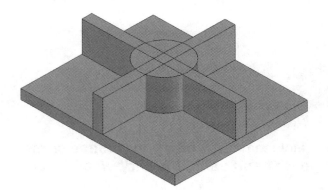

1 Cylinder and 3 Boxes need to be combined.

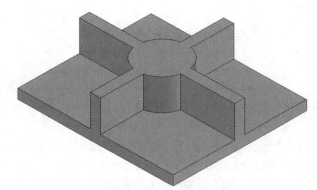

The 4 objects have now been combined into a single 3D solid model.

As you can see from the examples, complex 3D solid models can be created with ease by combing 2 or more 3D solid models using the **Union** tool. The possibilities are endless.

Union Tool....continued

There are various methods you can use to initiate the **Union** tool.

Method 1 – Solid Editing Panel on the 3D Tools Tab of the Drafting and Annotation Workspace.

1. Select the **Union** tool.

Method 2 – Boolean Panel on the Solid Tab of the 3D Modeling Workspace.

1. Select the **Union** tool.

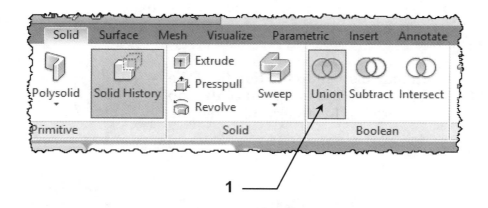

Method 3 – Keyboard entry.

1. On the **Command Line** or in the **Dynamic Input box**, type in *union* on your keyboard then press *<enter>*.

Union Tool....continued

An example of using the Union tool.

1. Create a 3D solid Box with the following sizes. You can place the Box anywhere on the screen. It doesn't have to be on 0,0,0.

<u>**Inch users**</u>

Length = 1.968"
Width = 0.984"
Height = 0.394"

<u>**Metric users**</u>

Length = 50 mm
Width   = 25 mm
Height = 10 mm

2. Create a 3D solid Cylinder with the following sizes, with the center point of the Cylinder at the midpoint of the bottom front edge. (Use the **Midpoint Object Snap** to place the center of the Cylinder.)

<u>**Inch users**</u>

Radius = 0.197"
Height = 0.394"

<u>**Metric users**</u>

Radius = 5 mm
Height = 10 mm

**You may also snap the height of the
Cylinder to the midpoint of the top
edge, or by entering the height.**

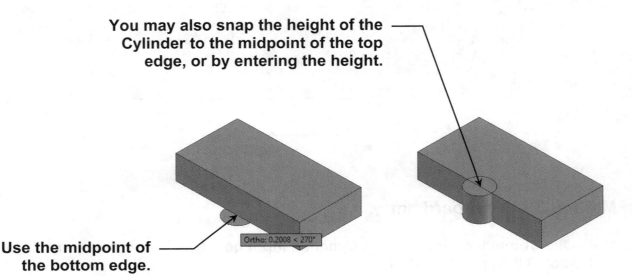

Ortho: 0.2008 < 270°

**Use the midpoint of
the bottom edge.**

Continued on the next page...

Union Tool....continued

Create 3 more identical 3D solid Cylinders by placing them on the midpoints of the other 3 top edges. The easiest way to achieve this is by using the **Copy** tool.

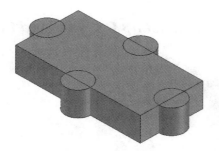

Selecting the Copy tool.

Method 1 – Modify Panel on the Home Tab of the Drafting and Annotation Workspace.

1. Select the **Copy** tool.

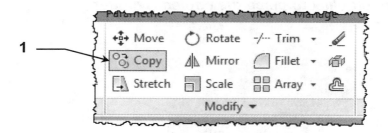

Method 2 – Modify Panel on the Home Tab of the 3D Modeling Workspace.

1. Select the **Copy** tool.

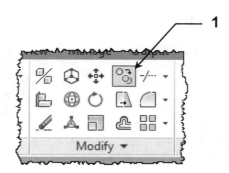

Method 3 – Keyboard entry.

1. On the **Command Line** or in the **Dynamic Input box**, type in *copy* on your keyboard then press *<enter>*.

Continued on the next page...

Union Tool....continued

Using the Copy tool for the 3 extra Cylinders.

1. Select the **Copy** tool.

2. Select objects: *left click on the existing Cylinder and then press <enter>.*

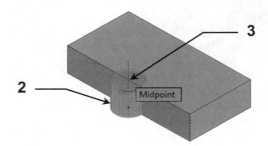

3. Specify base point or [**D**isplacement/m**O**de] <Displacement>: *left click on the center at the top of the Cylinder using the Midpoint Snap.*

4. Specify second point or [**A**rray]: <use first point as displacement>: *move the mouse cursor over to the midpoint of the top right-hand edge until you see the Midpoint Snap, then left click to place the copied Cylinder.*

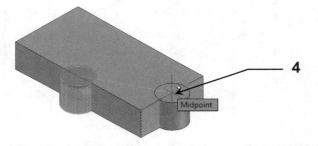

5. Specify second point or [**A**rray/**E**xit/**U**ndo] <Exit>: *move the mouse cursor over to the midpoint of the top back edge until you see the Midpoint Snap, then left click to place the copied Cylinder.*

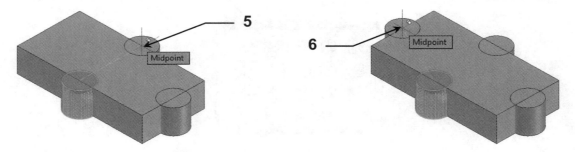

6. Specify second point or [**A**rray/**E**xit/**U**ndo] <Exit>: *repeat Step 5 for the final Cylinder on the midpoint of the top left-hand edge.*

7. Specify second point or [**A**rray/**E**xit/**U**ndo] <Exit>: *press <enter> to finish the Copy command.*

Continued on the next page...

Union Tool....continued

The 4 Cylinders have now been completed and need to be combined with the Box to create a single 3D solid model by using the **Union** command.

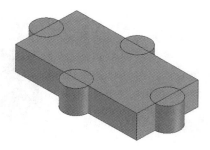

1. Select the **Union** command.

2. Select objects: *use a Window Selection or select them individually, and then press <enter>*.

3. The **Union** command will end automatically and all 5 solid models will be combined into one 3D solid model.

4. Save this completed model as **Union Exercise**. You will need it for the Subtract tool exercise.

Note: When copying objects such as the Cylinder to different locations, you may find it easier if you turn off **Ortho Mode** (*F8*). This enables you to move the mouse cursor around freely.

Subtract Tool

The **Subtract** tool allows you to subtract one 3D solid model from another 3D solid model. For example, you could subtract a Cylinder from a Box to create a hole in the Box. Some examples of using the Subtract tool are shown below.

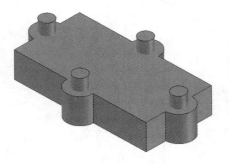

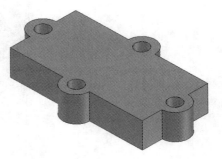

4 Cylinders need to be subtracted to create 4 holes.

4 holes have now been created after subtracting the Cylinders.

Note: The Cylinders have more height than the main shape. This makes it much easier to select the Cylinders during the Subtract operation.

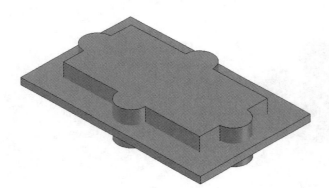

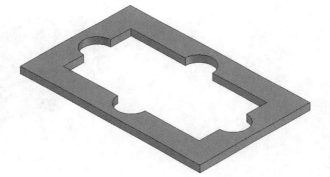

The inner model needs to be subtracted from the Box to create a hole.

The inner model has now been subtracted from the Box to create a hole.

As you can see from the examples, complex 3D solid models can be created with ease by subtracting 3D solid models from other 3D solid models using the **Subtract** tool.

Subtract Tool....continued

There are various methods you can use to initiate the **Subtract** tool.

Method 1 – Solid Editing Panel on the 3D Tools Tab of the Drafting and Annotation Workspace.

1. Select the **Subtract** tool.

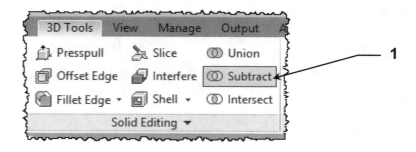

Method 2 – Boolean Panel on the Solid Tab of the 3D Modeling Workspace.

1. Select the **Subtract** tool.

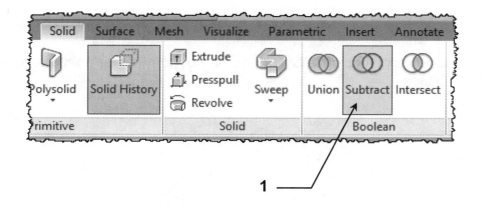

Method 3 – Keyboard entry.

1. On the **Command Line** or in the **Dynamic Input box**, type in *subtract* on your keyboard then press *<enter>*.

Subtract Tool....continued

An example of using the Subtract tool.

1. Open the **"Union Exercise"** drawing file if it is not already open.

2. Create a 3D solid Cylinder with the following sizes, with the center point of the Cylinder at the bottom center point of the half round lug on the front face. (Use the **Center Object Snap** to place the center of the Cylinder.)

Note: For this exercise it is easier to switch to **2D Wireframe** view when placing the Cylinder on the lug.

Inch users	Metric users
Radius = 0.098"	Radius = 2.5 mm
Height = 0.590"	Height = 15 mm

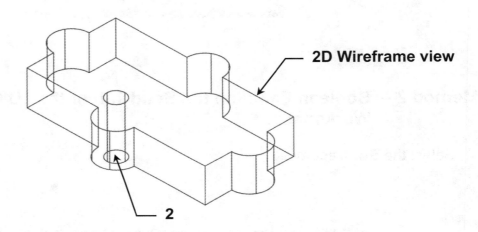

2D Wireframe view

2

3. Using the **Copy** tool, copy the Cylinder to the bottom center points on each of the remaining 3 lugs.

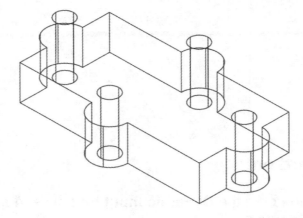

Continued on the next page...

Subtract Tool....continued

4. Select the **Union** tool.

5. Select objects: ***select the model with the 4 half round lugs and then press <enter>***. (Do not select the 4 Cylinders.)

6. Select objects: ***now select the 4 Cylinders and then press <enter>***.

7. The **Subtract** command will end automatically. You will notice that the 4 Cylinders have now reduced to the same height as the main model.

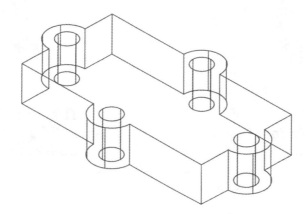

You may switch to Conceptual view for a more realistic look of the completed model. You could have completed this exercise while the model was in Conceptual view if you wished. When working with more complex models, I find it easier to work in 2D Wireframe view. It is a personal preference and the methods you choose to use are up to you.

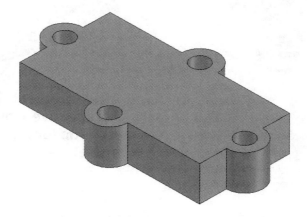

Save this completed model as **Union Exercise**.

Intersect Tool

The **Intersect** tool allows you to create a 3D solid model from overlapping 3D solid models that share the same space. Some examples of using the Intersect tool are shown below.

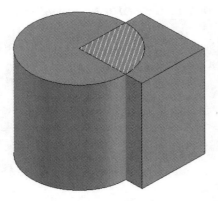

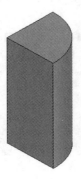

A 3D solid needs to be created from the Cylinder and Box that share the same space, shown hatched for clarity.

The completed 3D solid after the Intersect operation.

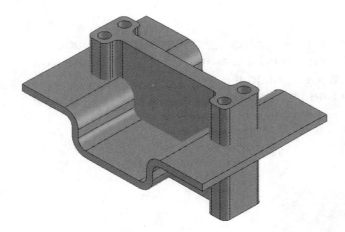

Before Intersect.

After Intersect. This is shown in Perspective view to give a more realistic look to the complex model.

As you can see from the examples, complex 3D solid models can be created with ease using the **Intersect** tool.

Intersect Tool....continued

There are various methods you can use to initiate the **Intersect** tool.

Method 1 – Solid Editing Panel on the 3D Tools Tab of the Drafting and Annotation Workspace.

1. Select the **Intersect** tool.

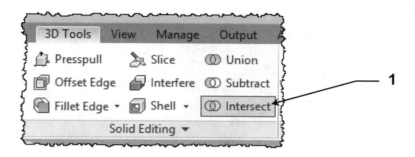

Method 2 – Boolean Panel on the Solid Tab of the 3D Modeling Workspace.

1. Select the **Intersect** tool.

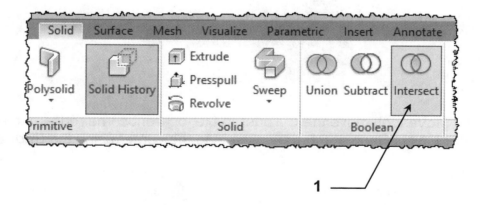

Method 3 – Keyboard entry.

1. On the **Command Line** or in the **Dynamic Input box**, type in *intersect* on your keyboard then press *<enter>*.

Intersect Tool....continued

An example of using the Intersect tool.

1. Create a 3D solid Box with the following sizes. You can place the Box anywhere on the screen. It doesn't have to be on 0,0,0.

Inch users

Length = 1.000"
Width = 1.000"
Height = 1.500"

Metric users

Length = 25.4 mm
Width = 25.4 mm
Height = 38.1 mm

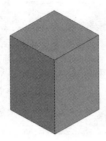

2. Create a 3D solid Cylinder with the following sizes. You can place the Cylinder anywhere on the screen. It doesn't have to be on 0,0,0.

Inch users

Radius = 0.750"
Height = 1.500"

Metric users

Radius = 19.05 mm
Height = 38.1 mm

3. Using the **Move** tool, move the Cylinder to the top front corner edge of the Box. Use the **Center** Object Snap at **P1**, then move it to the corner using the **Endpoint** Snap at **P2**.

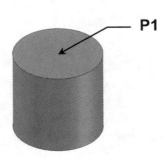 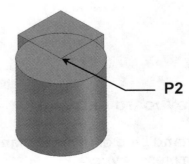 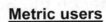

P1

P2

Continued on the next page...

Intersect Tool....continued

4. Select the **Intersect** tool.

5. Select objects: *select the Box and the Cylinder and then press <enter>.*

6. The **Intersect** command will end automatically and a new 3D solid model will be created by the shared space that the Box and Cylinder occupied.

Exercise 5A

1. Start a new drawing file by selecting either **acad.dwt** for inch users, or **acadiso.dwt** for metric users.

2. Select the **SE Isometric** view.

3. Save the drawing file as **Ex-5A** but keep the drawing file open.

4. Create 2 Boxes with the following sizes. You can place them anywhere in the drawing area as they will be moved into position in the next step.

Box 1

Inch users

Length = 1.500"
Width   = 1.500"
Height = 0.750"

Metric users

Length = 38.1 mm
Width   = 38.1 mm
Height = 19.05 mm

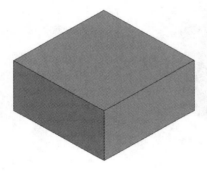

Box 2

Inch users

Length = 0.750"
Width   = 1.500"
Height = 0.750"

Metric users

Length = 19.05 mm
Width   = 38.1 mm
Height = 19.05 mm

Continued on the next page...

Exercise 5A....continued

5. Move the smaller Box on top of the larger Box so the two boxes are inline as shown.

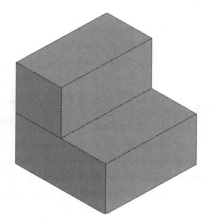

6. Resave the drawing file as **Ex-5A** but keep the drawing file open.

7. Create 2 Cylinders with the following sizes. You can place them anywhere in the drawing area as they will be moved into position in the next step.

Cylinder 1

Inch users

Radius = 0.400"
Height = 0.750"

Metric users

Radius = 10.16 mm
Height = 19.05 mm

Cylinder 2

Inch users

Radius = 0.250"
Height = 0.750"

Metric users

Radius = 6.35 mm
Height = 19.05 mm

Continued on the next page...

Exercise 5A....continued

8. Move the 2 Cylinders so the center of each Cylinder is on the midpoints of the edges as shown.

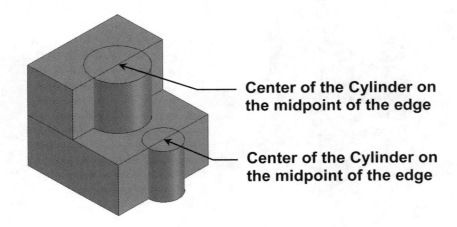

Center of the Cylinder on the midpoint of the edge

Center of the Cylinder on the midpoint of the edge

9. Resave the drawing file as **Ex-5A** but keep the drawing file open.

10. Use the **Union** tool to join all 4 of the 3D solids into one 3D solid.

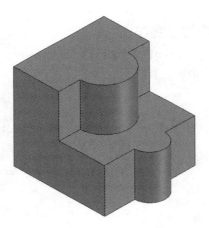

11. Resave the drawing file as **Ex-5A**.

Exercise 5B

1. Open **Ex-5A**.

2. Create 2 Cylinders with the following sizes. You can place them anywhere in the drawing area as they will be moved into position in the next step.

Cylinder 1

Inch users

Radius = 0.250"
Height = 1.000"

Metric users

Radius = 6.35 mm
Height = 25.4 mm

Cylinder 2

Inch users

Radius = 0.125"
Height = 1.000"

Metric users

Radius = 3.17 mm
Height = 25.4 mm

3. Save the drawing file as **Ex-5B** but keep the drawing file open.

The 2 Cylinders now need to be placed on the center points of the lugs. You will need to override certain Object Snaps and use just the **Center Snap** for this operation. The instructions on how to achieve this are on the next page.

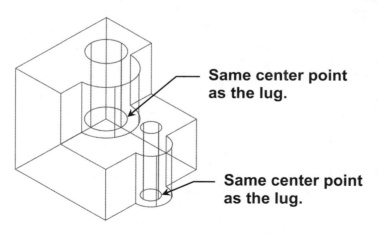

Same center point as the lug.

Same center point as the lug.

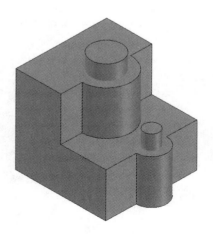

Continued on the next page...

Exercise 5B....continued

1. Select the **Move** tool.

2. Select objects: *left click on the large Cylinder and then press <enter>.*

3. Specify base point or [**D**isplacement] <Displacement>: *using the Center Object Snap, left click on the center point at the bottom of the Cylinder (P1).*

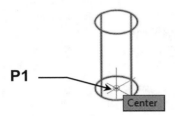

4. Specify second point or <use first point as displacement>: *hold down the <u>Shift</u> key on your keyboard and then right click with your mouse.*

5. Select **Center** from the list.

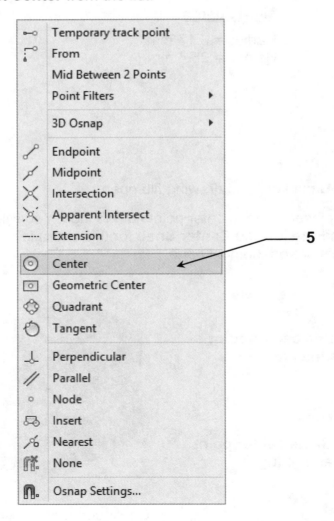

Continued on the next page...

Exercise 5B....continued

6. Specify second point or <use first point as displacement>: *drag your mouse cursor over to the bottom center point of the large upper lug and using the Center Object Snap, left click on the center point (P2).*

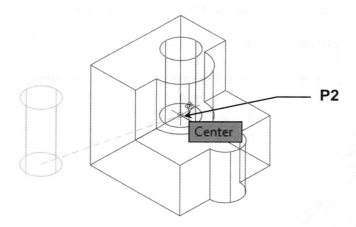

7. Repeat **Steps 1-6** for the smaller Cylinder, placing it on the lower lug.

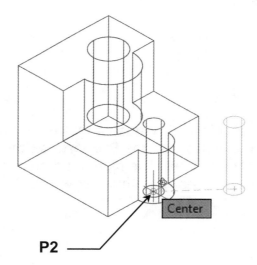

8. Resave the drawing file as **Ex-5B** but keep the drawing file open.

Note: Using the *Shift* key and right mouse click to select an **Object Snap** from the list, is an efficient way of overriding any other Object Snaps that you may have set in the **Drafting Settings** dialog box. For example, the **Endpoint** and **Midpoint** Object Snaps.

Continued on the next page...

Exercise 5B....continued

The 2 Cylinders now need to be subtracted from the main 3D solid model.

1. Select the **Subtract** tool.

2. Select objects: ***select the main 3D solid model and then press <enter>.***

3. Select objects: ***select the 2 Cylinders and then press <enter>.***

4. The **Subtract** command will end automatically and the model will be complete.

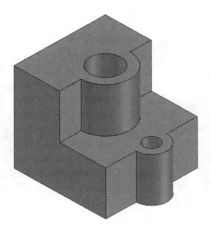

5. Resave the drawing file as **Ex-5B**.

The image below shows the 3D solid model with a Transparency of **70** applied to it. To achieve this simply select the model and then select the **Properties Palette**. In the **Transparency** field enter a number between **0-90**. The higher the number the more transparent the model will be. For no transparency the number will be **0** (zero).

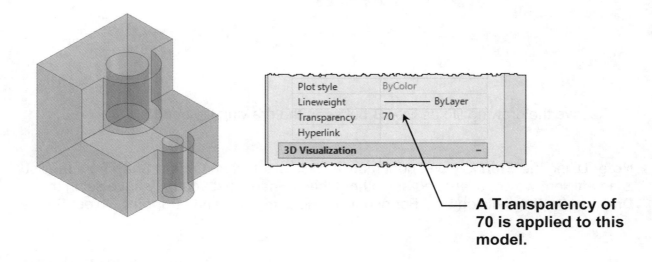

A Transparency of 70 is applied to this model.

Exercise 5C

1. Start a new drawing file by selecting either **acad.dwt** for inch users, or **acadiso.dwt** for metric users.

2. Select the **SE Isometric** view.

3. Save the drawing file as **Ex-5C** but keep the drawing file open.

4. Create a Box with the following sizes. You can place the Box anywhere in the drawing area.

    Inch users

    Length = 2.362"
    Width = 0.236"
    Height = 1.771"

    Metric users

    Length = 60 mm
    Width = 6 mm
    Height = 45 mm

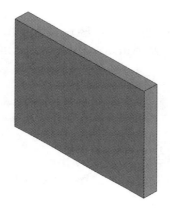

5. Create a Truncated Cone with the following sizes. You can place the Cone anywhere in the drawing area.

    Inch users

    Base radius = 0.748"
    Top radius = 0.236"
    Height = 1.771"

    Metric users

    Base radius = 19 mm
    Top radius = 6 mm
    Height = 45 mm

Continued on the next page...

Exercise 5C....continued

6. Create a line from the midpoint of the top front edge of the Box to the midpoint of the top back edge of the Box. (Use the **Midpoint** Object Snap.)

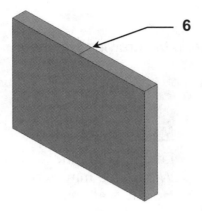

7. Move the Truncated Cone over to the center of the Box using the **Center** Snap on the top radius of the Cone, and using the **Midpoint** Snap to position the Cone on the midpoint of the line as shown.

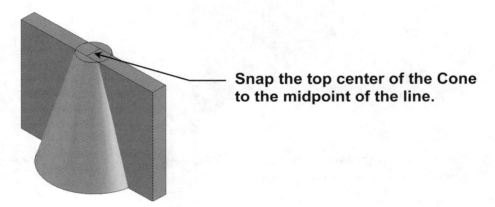

Snap the top center of the Cone to the midpoint of the line.

8. Using the **Intersect** tool, create a 3D solid model from the space that both models occupy.

9. Delete the line on the top face.

10. Resave the drawing file as **Ex-5C**.

LEARNING OBJECTIVES

After completing this lesson you will be able to:

1. Rotate 3D Models.
2. Create a Mirrored Copy of 3D Models.
3. Align 3D Models.

LESSON 6

3D Rotate Tool

The **3D Rotate** tool allows you to rotate a 3D solid model about the X, Y or Z axis. You first pick a base point for the rotation and then pick one of the 3 colored ribbons (shown below) for the axis of rotation.

Red ribbon: Rotates the model around the **X axis**.

Green ribbon: Rotates the model around the **Y axis**.

Blue ribbon: Rotates the model around the **Z axis**.

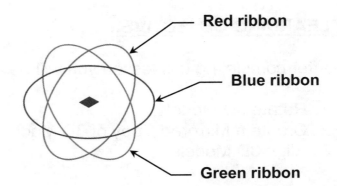

— **Red ribbon**

— **Blue ribbon**

— **Green ribbon**

There are various methods you can use to initiate the **3D Rotate** tool.

Method 1 – Modify Panel on the Home Tab of the 3D Modeling Workspace.

1. Select the **3D Rotate** tool.

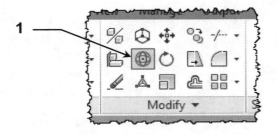

Method 2 – Keyboard entry.

1. On the **Command Line** or in the **Dynamic Input box**, type in *3drotate* on your keyboard then press *<enter>*.

3D Rotate Tool....continued

An example of rotating a 3D solid model 90 degrees around the X axis.

1. Select the **3D Rotate** tool.

2. Select objects: *left click on the 3D solid model and then press <enter>.*

3. Specify base point: *left click on the model to select a base point (P1).*

4. Pick a rotation axis: *left click on the red ribbon (P2).*

5. Specify angle start point or type an angle: *type in 90 and then press <enter>.*

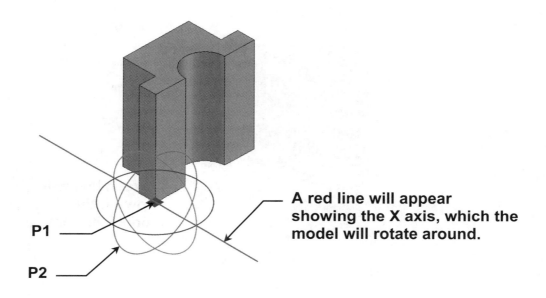

P1 ——

P2 ——

— A red line will appear
showing the X axis, which the
model will rotate around.

6. The 3D solid model will rotate 90 degrees around the X axis and the **3D Rotate** command will end automatically.

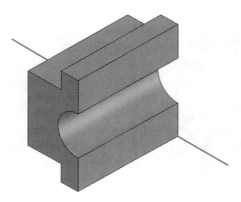

3D Rotate Tool....continued

An example of rotating a 3D solid model 90 degrees around the Y axis.

1. Select the **3D Rotate** tool.

2. Select objects: ***left click on the 3D solid model and then press <enter>.***

3. Specify base point: ***left click on the model to select a base point (P1).***

4. Pick a rotation axis: ***left click on the green ribbon (P2).***

5. Specify angle start point or type an angle: ***type in 90 and then press <enter>.***

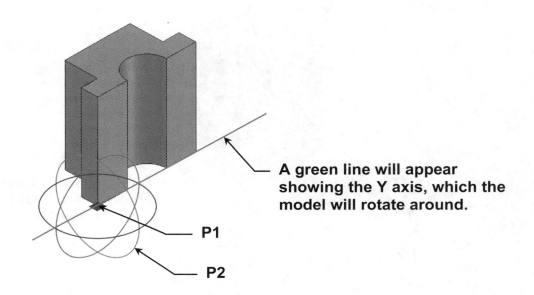

A green line will appear showing the Y axis, which the model will rotate around.

P1

P2

6. The 3D solid model will rotate 90 degrees around the Y axis and the **3D Rotate** command will end automatically.

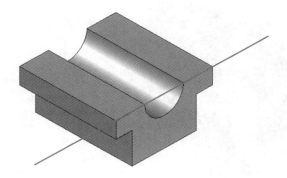

3D Rotate Tool....continued

An example of rotating a 3D solid model 90 degrees around the Z axis.

1. Select the **3D Rotate** tool.

2. Select objects: *left click on the 3D solid model and then press <enter>.*

3. Specify base point: *left click on the model to select a base point (P1).*

4. Pick a rotation axis: *left click on the blue ribbon (P2).*

5. Specify angle start point or type an angle: *type in 90 and then press <enter>.*

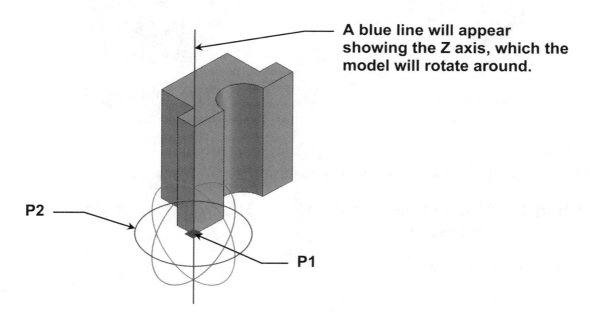

A blue line will appear showing the Z axis, which the model will rotate around.

P2

P1

6. The 3D solid model will rotate 90 degrees around the Z axis and the **3D Rotate** command will end automatically.

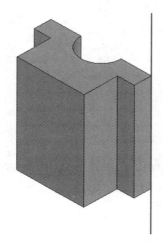

3D Mirror Tool

The **3D Mirror** tool allows you to create a mirrored copy of a 3D solid model by selecting a mirror plane. The most commonly used method is by selecting **3 points** as the mirror plane.

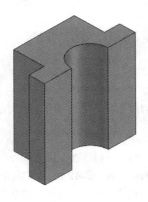

Before mirroring.

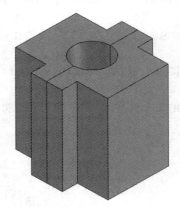

After mirroring.

There are various methods you can use to initiate the **3D Mirror** tool.

Method 1 – Modify Panel on the Home Tab of the 3D Modeling Workspace.

1. Select the **3D Mirror** tool.

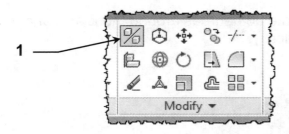

Method 2 – Keyboard entry.

1. On the **Command Line** or in the **Dynamic Input box**, type in *mirror3d* on your keyboard then press *<enter>*.

3D Mirror Tool....continued

An example of mirroring a 3D solid model using 3 points for the mirror plane.

1. Select the **3D Mirror** tool.

2. Select objects: *left click on the 3D solid model and then press <enter>.*

3. Specify first point of mirror plane (3 points) or [Object/Last/**Z**axis/**V**iew/**XY/YZ/ZX**/3points] <3points>: *type in 3 and then press <enter>.*

4. Specify first point on mirror plane: *left click using the Endpoint Snap at (P1).*

5. Specify second point on mirror plane: *left click using the Endpoint Snap at (P2).*

6. Specify third point on mirror plane: *left click using the Endpoint Snap at (P3).*

7. Delete source objects? [**Y**es/**N**o] <N>: *type in n* (for No) *and then press <enter>.*

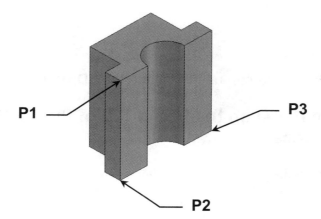

8. A mirrored copy will be created and the **3D Mirror** command will end automatically.

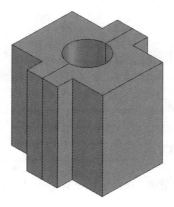

Note: If you had entered *y* (for Yes) at **Step 7** a mirrored copy would have been created but the original 3D solid model (source object) would have been deleted.

3D Align Tool

The **3D Align** tool allows you to move and rotate a 3D solid model by selecting up to 3 source points and then 3 destination points. In the example below a small Box needs to be moved and rotated to align with the top angled part of the larger Box.

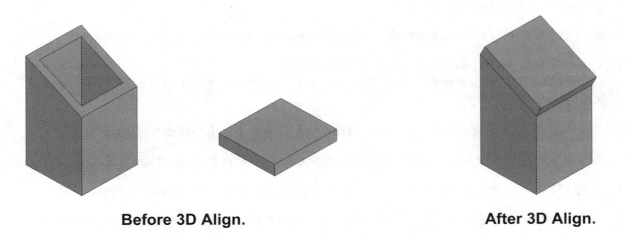

Before 3D Align.   **After 3D Align.**

There are various methods you can use to initiate the **3D Align** tool.

Method 1 – Modify Panel on the Home Tab of the 3D Modeling Workspace.

1. Select the **3D Align** tool.

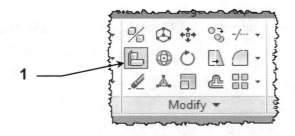

Method 2 – Keyboard entry.

1. On the **Command Line** or in the **Dynamic Input box**, type in *3dalign* on your keyboard then press *<enter>*.

3D Align Tool....continued

An example of aligning a 3D solid model using 3 source points and 3 destination points.

1. Select the **3D Align** tool.

2. Select objects: *left click on the object to be aligned and then press <enter>.*

   Specify source plane and orientation …

3. Specify base point or [**C**opy]: *left click on the first source point (S1).*

4. Specify second point or [**C**ontinue] <C>: *left click on the second source point (S2).*

5. Specify third point or [**C**ontinue] <C>: *left click on the third source point (S3).*

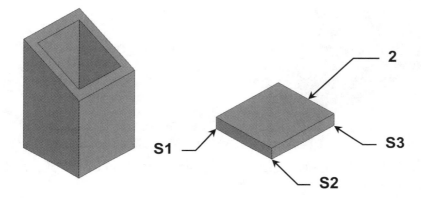

   Specify destination plane and orientation …

6. Specify first destination point: *left click on the first destination point (D1).*

7. Specify second destination point or [e**X**it] <X>: *left click on the second destination point (D2).*

8. Specify third destination point or [e**X**it] <X>: *left click on the third destination point (D3).*

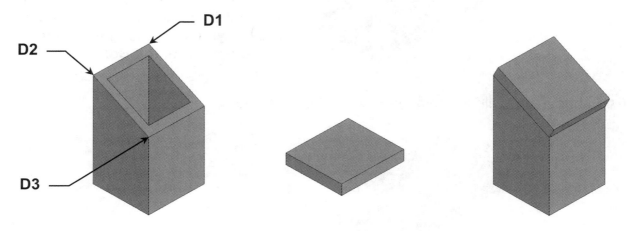

9. The source object has now been moved and rotated to align with the top angled part of the larger Box and the **3D Align** command will end automatically.

Exercise 6A

1. Start a new drawing file by selecting either **acad.dwt** for inch users, or **acadiso.dwt** for metric users.

2. Select the **SE Isometric** view.

3. Save the drawing file as **Ex-6A** but keep the drawing file open.

4. Create a Box with the following sizes.

<u>**Inch users**</u>

Length = 2.000"
Width = 1.000"
Height = 0.250"

<u>**Metric users**</u>

Length = 50.8 mm
Width = 25.4 mm
Height = 6.35 mm

5. Create a Cylinder with the following sizes and place it in the center of the Box as shown.

<u>**Inch users**</u>

Radius = 0.250"
Height = 0.500"

<u>**Metric users**</u>

Radius = 6.35 mm
Height = 12.7 mm

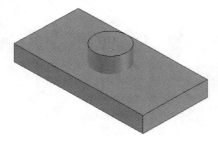

Front transparent view showing the Cylinder placed level with the bottom of the Box.

6. Subtract the Cylinder from the Box to create a hole.

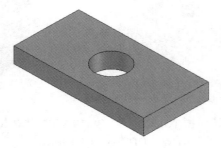

Continued on the next page...

Exercise 6A....continued

7. Rotate the 3D model by 90 degrees in the **X** axis.

8. Create a Wedge with the following sizes.

Inch users

    Length = 0.500"
    Width   = 0.250"
    Height = 0.500"

Metric users

    Length = 12.7 mm
    Width   = 6.35 mm
    Height = 12.7 mm

9. Move the Wedge to the top right-hand end of the larger 3D model as shown.

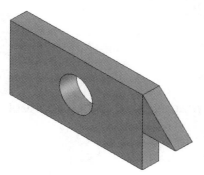

10. Join both models together to form one model using the **Union** tool.

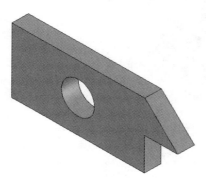

Continued on the next page...

Exercise 6A....continued

11. Create a Box with the following sizes.

Inch users

    Length = 2.000"
    Width = 1.000"
    Height = 0.250"

Metric users

    Length = 50.8 mm
    Width = 25.4 mm
    Height = 6.35 mm

12. Position the Box under the 3D model as shown.

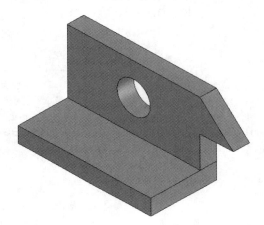

13. Join both models together to form one model using the **Union** tool.

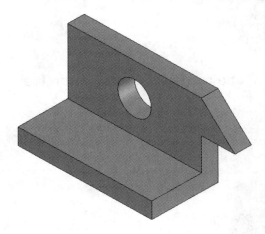

14. Resave the drawing file as **Ex-6A**.

Continued on the next page...

Exercise 6B

1. Open **Ex-6A** if not already open.

2. Using the **3D Mirror** tool, mirror the 3D model using the 3 points as shown.

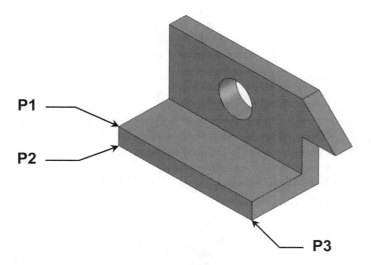

3. Join both models together to form one model using the **Union** tool.

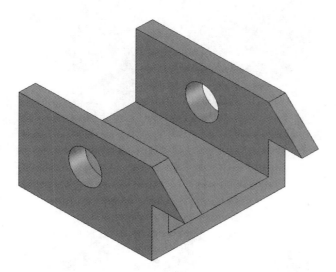

4. Save the drawing file as **Ex-6B** but keep the drawing file open.

Continued on the next page...

Exercise 6B....continued

5. Create a Box with the following sizes.

Inch users

Length = 2.000"
Width   = 0.250"
Height = 0.125"

Metric users

Length = 50.8 mm
Width   = 6.35 mm
Height = 3.17 mm

6. Using the **3D Align** tool, align the 3D Box to the main 3D model using the 3 source points and the 3 destination points as shown.

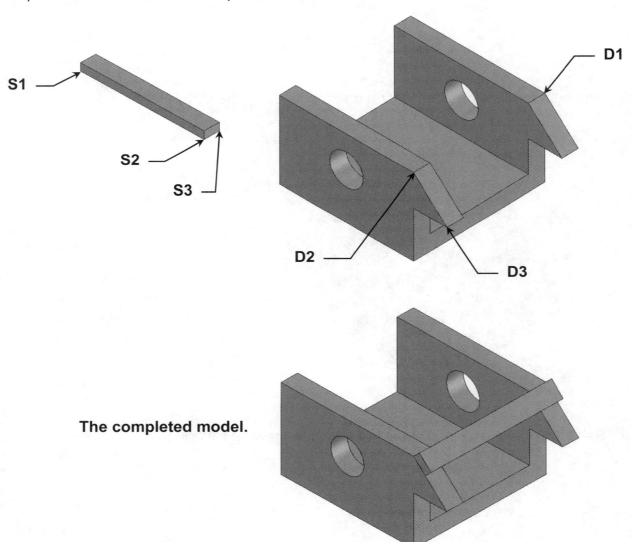

The completed model.

7. Resave the drawing file as **Ex-6B**.

LEARNING OBJECTIVES

After completing this lesson you will be able to:

1. Extrude Complex 2D Profiles into 3D Solid Models.
2. Revolve Complex 2D Profiles into 3D Solid Models.
3. Loft 2D Cross Sections to Create Complex 3D Solid Models.
4. Sweep 2D Profiles along a Path to Create 3D Solid Models.

LESSON 7

Extrude Tool

So far you have been using AutoCAD's basic 3D tools to create 3D solid models. There will be times when you need to create more complex 3D models. This is achieved by first creating a 2D closed shape and then using the **Extrude** tool to extrude the 2D shape into a 3D solid model. You can also have tapered sides on the extruded model. Some examples of extruding 2D shapes into 3D solid models are shown below.

It is assumed that you are already familiar with 2D drafting using AutoCAD, but there are sample files for you to download that will aid in the exercises for this lesson. (See page Intro-1 in the Introduction to this Workbook for instructions on how to download the sample files.)

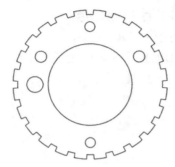

Plan view of the 2D shape before extruding.

SE Isometric view after using the Extrude tool and subtracting the holes.

Plan view of the 2D shape before extruding.

Perspective view after using the Extrude tool and then subtracting the holes.

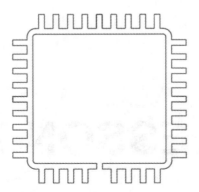

Plan view of the 2D shape before extruding.

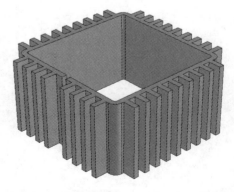

Perspective view after using the Extrude tool.

Extrude Tool....continued

There are various methods you can use to initiate the **Extrude** tool.

Method 1 – Modeling Panel on the 3D Tools Tab of the Drafting and Annotation Workspace.

1. Select the **Extrude** tool.

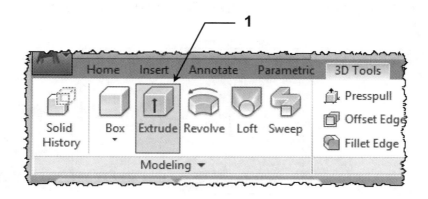

Method 2 – Solid Panel on the Solid Tab of the 3D Modeling Workspace.

1. Select the **Extrude** tool.

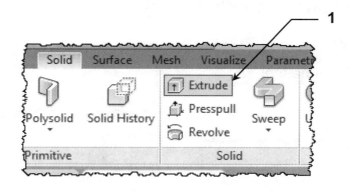

Method 3 – Keyboard entry.

1. On the **Command Line** or in the **Dynamic Input box**, type in *extrude* on your keyboard then press *<enter>*.

Extrude Tool....continued

An example of using the Extrude tool.

Open the sample drawing file **"Square Hollow Section - Inch.dwg"** if you are an inch user, or **"Square Hollow Section - Metric.dwg"** if you are a metric user.

This drawing file is a simple 2D plan view of a square hollow section. Both the inner and outer squares need to be extruded in height. The next stage is to subtract the inner square from the outer square to create a 3D solid model of a square hollow section.

All the lines and arcs on the outer and inner squares are separate objects and need to be joined together to create 2 closed shapes. This is achieved by using the **Edit Polyline** tool.

There are various methods you can use to initiate the **Edit Polyline** tool.

Method 1 – Modify drop-down Panel on the Home Tab of the Drafting and Annotation Workspace.

1. Select the Modify drop-down Panel.

2. Select the **Edit Polyline** tool.

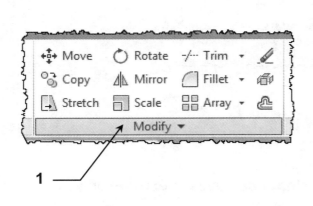

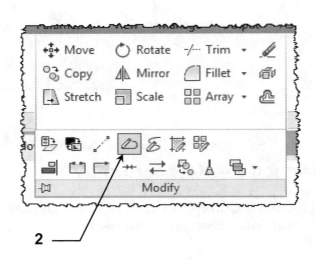

Extrude Tool....continued

Method 2 – Modify drop-down Panel on the Home Tab of the 3D Modeling Workspace.

1. Select the Modify drop-down Panel.

2. Select the **Edit Polyline** tool.

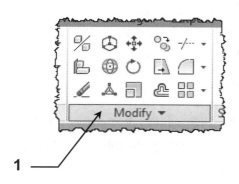

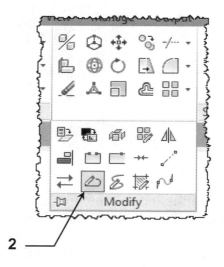

1

2

Method 3 – Keyboard entry.

1. On the **Command Line** or in the **Dynamic Input box**, type in *pedit* on your keyboard then press *<enter>*.

How to create the 2 closed shapes using the Edit Polyline tool.

1. Select the **Edit Polyline** tool.

2. Select Polyline or [**M**ultiple]: *select one of the outer lines.*

3. Do you want to turn it into one? <Y> *type in y* (for Yes) *and then press <enter>.*

4. Enter an option [**C**lose/**J**oin/**W**idth/**E**dit vertex/**F**it/**S**pline/**D**ecurve/**L**type gen/**R**everse/**U**ndo]: *type in j* (for Join) *and then press <enter>.*

Note: You can also select the **Join** option from the Dynamic Input Menu.

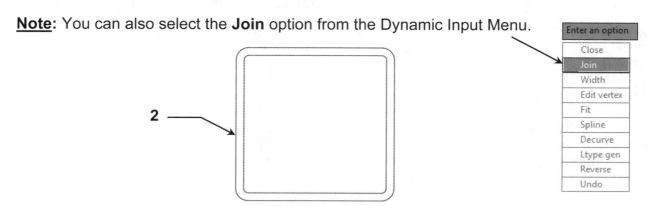

Continued on the next page...

Extrude Tool....continued

5. Select objects: *use a Window selection to select all the outer lines and arcs, and then press <enter>.*

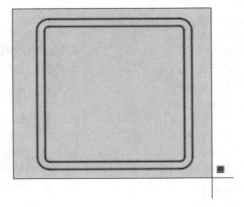

Note: Although the inner lines and arcs will be selected as well, only the outer lines and arcs will be joined into one object.

6. Enter an option [**C**lose/**J**oin/**W**idth/**E**dit vertex/**F**it/**S**pline/**D**ecurve/**L**type gen/**R**everse/ **U**ndo]: *press <enter> again to end the command.*

If you hover the mouse cursor over one of the outer lines you will see the outside square highlight to show you it is now a closed Polyline.

The outside lines and arcs have now become one closed Polyline.

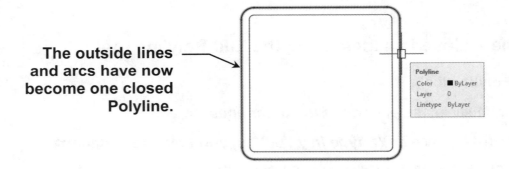

7. Now repeat **Steps 2 - 6** for the inner lines and arcs, making sure you select an inner line at **Step 2**.

The inner lines and arcs have now become one closed Polyline.

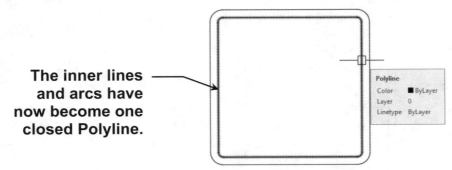

Continued on the next page...

Extrude Tool....continued

Extruding the 2 closed Polyline Squares.

1. Select the **SE Isometric** view.

2. Switch to **Conceptual** style.

3. Select the **Extrude** tool.

4. Select objects to extrude or [Mode]: *select the outer and inner closed polylines and then press <enter>.*

5. Specify height of extrusion or [Direction/**P**ath/**T**aper angle/**E**xpression] <0>: *move your mouse upward then type in 2.5 [65] and then press <enter>.*

6. The two closed Polylines will now be extruded to the height you specified and into 2 separate 3D solid models and the **Extrude** command will end automatically.

7. Use the **Subtract** tool to subtract the inner 3D solid model from the outer 3D solid model.

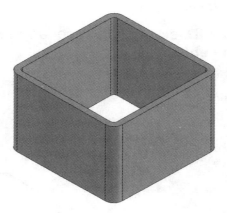

You have now created a 3D solid model of a **Square Hollow Section** from 2 separate 2D shapes. Save the drawing file as **Square Hollow Section 3D**.

Extrude Tool....continued

An example of using the Extrude tool with tapered sides.

Open the sample drawing file **"Tapered Block - Inch.dwg"** if you are an inch user, or **"Tapered Block - Metric.dwg"** if you are a metric user.

This drawing file is a simple 2D plan view of a shaped block that needs to be extruded in height and have tapered sides with a 10 degree angle applied to them to create a 3D solid model.

1. Join all the lines and arcs into a closed shape using the **Edit Polyline** tool.

2. Select the **SE Isometric** view.

3. Switch to **Conceptual** style.

4. Select the **Extrude** tool.

5. Select objects to extrude or [**M**ode]: *select the closed Polyline shape and then press <enter>.*

6. Move your mouse up until you see the model gain height

7. Specify height of extrusion or [**D**irection/**P**ath/**T**aper angle/**E**xpression] <0>: *type in t* (for Taper angle) *and then press <enter>.*

8. Specify angle of taper for extrusion or [**E**xpression]: <0> *type in 10* (for the Taper angle) *and then press <enter>.*

9. Specify height of extrusion or [**D**irection/**P**ath/**T**aper angle/**E**xpression] <0>: *type in 1.378 [35] and then press <enter>.*

10. The 3D tapered model with tapered sides is now complete and the **Extrude** command will end automatically.

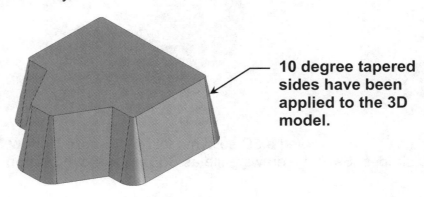

10 degree tapered sides have been applied to the 3D model.

Revolve Tool

The **Revolve** tool allows you to revolve a 2D profile around an axis. You can revolve the profile around any axis from 1 to 360 degrees. Some examples of revolving 2D shapes into 3D solid models are shown below.

It is assumed that you are already familiar with 2D drafting using AutoCAD, but there are sample files for you to download that will aid in the exercises for this lesson. (See page Intro-1 in the Introduction to this Workbook for instructions on how to download the sample files.)

Plan view of the 2D profile before revolving.

Front view of the 3D model after revolving.

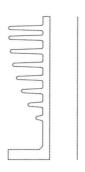

Plan view of the 2D profile before revolving around a central line.

Perspective view of the 3D model after revolving.

Plan view of the 2D profile before revolving.

Front view of the 3D model after revolving.

As you can see from the examples, complex 3D solid models can be created with ease by first creating the 2D profile and then using the **Revolve** tool.

Revolve Tool....continued

There are various methods you can use to initiate the **Revolve** tool.

Method 1 – Modeling Panel on the 3D Tools Tab of the Drafting and Annotation Workspace.

1. Select the **Revolve** tool.

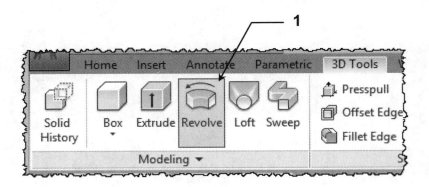

Method 2 – Solid Panel on the Solid Tab of the 3D Modeling Workspace.

1. Select the **Revolve** tool.

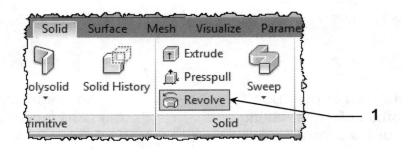

Method 3 – Keyboard entry.

1. On the **Command Line** or in the **Dynamic Input box**, type in *revolve* on your keyboard then press *<enter>*.

Revolve Tool....continued

An example of using the Revolve tool.

Open the sample drawing file **"Piston Barrel - Inch.dwg"** if you are an inch user, or **"Piston Barrel - Metric.dwg"** if you are a metric user.

This drawing file is a simple 2D plan view of a piston barrel that needs to be revolved 360 degrees around a central line to create a 3D solid model.

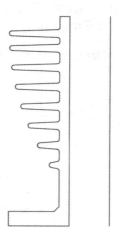

All the lines and arcs are separate objects and need to be joined together to create a closed profile. This is achieved by using the **Edit Polyline** tool you used in the Extrude example. **Note:** The central line is not part of the profile and is just used as the central point to revolve the profile around.

When selecting the lines and arcs during the **Edit Polyline** command, make sure you do not include the central line in the Window selection as shown below.

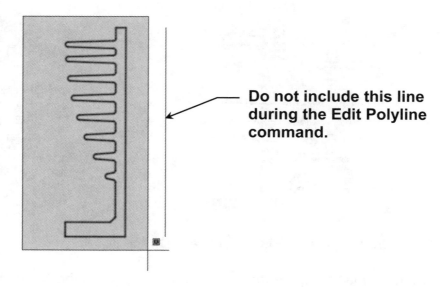

Do not include this line during the Edit Polyline command.

Continued on the next page...

Revolve Tool....continued

Revolving the closed profile.

1. Switch to **Conceptual** style.

2. Select the **Revolve** tool.

3. Select objects to revolve or [Mode]: *select the closed profile and then press <enter>.*

4. Specify axis start point or define axis by [Object/X/Y/Z] <Object>: *left click at the top of the single line using the Endpoint Snap (P1).*

5. Specify axis endpoint: *left click at the bottom of the single line using the Endpoint Snap (P2).*

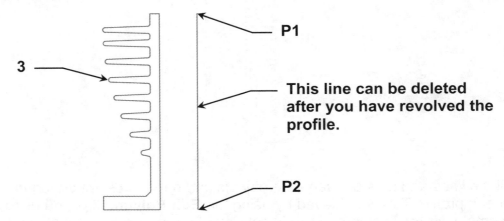

P1

3

This line can be deleted after you have revolved the profile.

P2

6. Specify angle of revolution or [Start angle/Reverse/Expression]: *type in 360 and then press <enter.*

7. The 2D profile will have revolved about the central line 360 degrees and the **Revolve** command will end automatically.

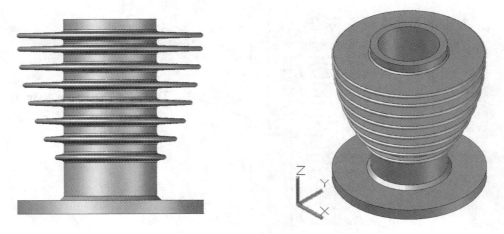

8. Save the drawing file as **Piston Barrel 3D**.

Note: You may wish to rotate the 3D solid model around the X axis by 90 degrees positive.

Loft Tool

The **Loft** tool allows you to create complex 3D solid models from a series of 2D cross sections. Some examples of lofting 2D cross sections into 3D solid models are shown below.

It is assumed that you are already familiar with 2D drafting using AutoCAD, but there are sample files for you to download that will aid in the exercises for this lesson. (See page Intro-1 in the Introduction to this Workbook for instructions on how to download the sample files.)

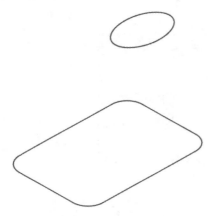

SE Isometric view of rectangular cross section on the Z zero plane, and an elliptical cross section projected in the Z plus axis.

The finished 3D solid model after using the Loft tool and selecting both cross sections.

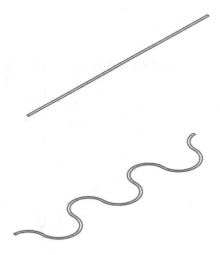

NE Isometric view of a series of arcs joined as a cross section on the Z zero plane, and a rectangular cross section projected in the Z plus axis.

The finished 3D solid model after using the Loft tool and selecting both cross sections.

As you can see from the examples, complex 3D solid models can be created with ease by first creating the 2D cross sections and then using the **Loft** tool.

Loft Tool....continued

There are various methods you can use to initiate the **Loft** tool.

Method 1 – Modeling Panel on the 3D Tools Tab of the Drafting and Annotation Workspace.

1. Select the **Loft** tool.

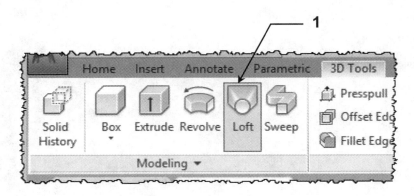

Method 2 – Sweep drop-down list on the Solid Panel of the Solid Tab of the 3D Modeling Workspace.

1. Select the **Sweep** drop-down list.

2. Select the **Loft** tool.

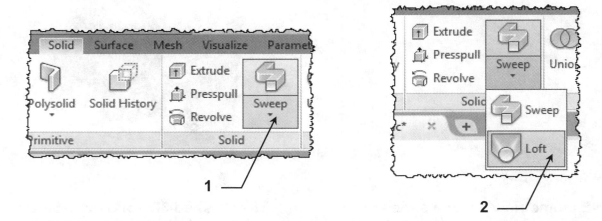

Note: AutoCAD remembers the last tool you used in a drop-down list.

Method 3 – Keyboard entry.

1. On the **Command Line** or in the **Dynamic Input box**, type in *loft* on your keyboard then press *<enter>*.

Loft Tool....continued

An example of using the Loft tool.

Open the sample drawing file **"Lofted Block - Inch.dwg"** if you are an inch user, or **"Lofted Block - Metric.dwg"** if you are a metric user.

This drawing file is a 2D SE Isometric view of a rectangular lower cross section and an elliptical upper cross section that need to be joined together using the **Loft** tool to create a 3D solid model.

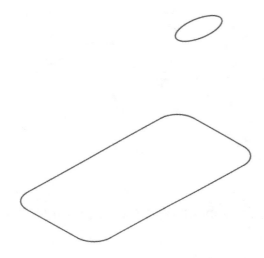

All the lines and arcs in the lower rectangular cross section have already been joined together so there is no need to use the **Edit Polyline** tool for this exercise.

Lofting the 2 cross sections.

1. Switch to **Conceptual** style.

2. Select the **Loft** tool.

Continued on the next page...

Loft Tool....continued

3. Select objects cross sections in lofting order or [**PO**int/**J**oin multiple edges/**MO**de]: *left click on the lower rectangular cross section (P1).*

4. Select objects cross sections in lofting order or [**PO**int/**J**oin multiple edges/**MO**de]: *left click on the upper elliptical cross section (P2).*

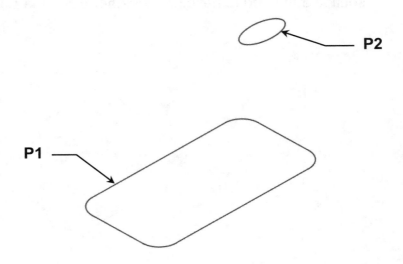

5. Select objects cross sections in lofting order or [**PO**int/**J**oin multiple edges/**MO**de]: *press <enter>.*

6. Enter an option [**G**uides/**P**ath/**C**ross section only/**S**ettings] <Cross section only>: *press <enter> again to end the Loft command.*

The completed 3D solid model after using the **Loft** tool.

7. Save the drawing file as **Lofted Block 3D**.

Sweep Tool

The **Sweep** tool allows you to create complex 3D solid models by extruding a 2D profile along a path. Some examples of sweeping 2D profiles into 3D solid models are shown below.

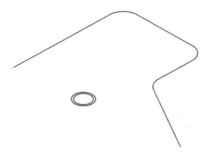

SE Isometric view of 2 circles that need to be swept along the polyline path to create a tube.

The finished 3D solid model after using the Sweep tool and subtracting the inner model from the outer model.

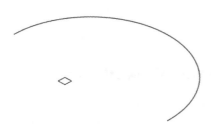

SE Isometric view of a rectangle that needs to be swept along the semi-circular path with a twist of 360 degrees.

The finished 3D solid model after using the Sweep tool and applying a twist of 360 degrees.

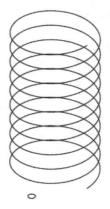

SE Isometric view of a circle that needs to be swept along the helix path to form a spring.

The finished 3D solid model in perspective view after using the Sweep tool.

As you can see from the examples, complex 3D solid models can be created with ease by first creating the 2D profile and a path, and then using the **Sweep** tool.

Sweep Tool....continued

There are various methods you can use to initiate the **Sweep** tool.

Method 1 – Modeling Panel on the 3D Tools Tab of the Drafting and Annotation Workspace.

1. Select the **Sweep** tool.

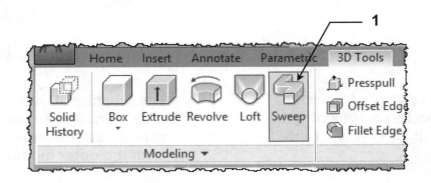

Method 2 – Solid Panel of the Solid Tab of the 3D Modeling Workspace.

1. Select the **Sweep** tool.

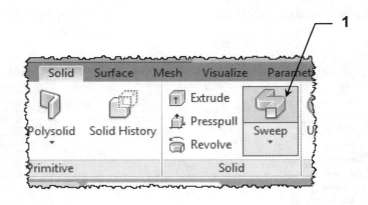

Method 3 – Keyboard entry.

1. On the **Command Line** or in the **Dynamic Input box**, type in *sweep* on your keyboard then press *<enter>*.

Sweep Tool....continued

An example of using the Sweep tool.

Open the sample drawing file **"Pipe - Inch.dwg"** if you are an inch user, or **"Pipe - Metric.dwg"** if you are a metric user.

This drawing file is a 2D SE Isometric view of a 2D path and 2 circles that need to be swept along the 2D path to create 2 3D solid models. The inner model is then subtracted from the outer model to create a tube.

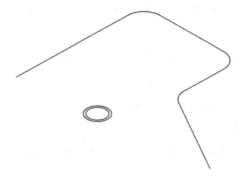

All the lines and arcs in the 2D path have already been joined together so there is no need to use the **Edit Polyline** tool for this exercise.

Sweeping the circle along the 2D path.

1. Switch to **Conceptual** style.

2. Select the **Sweep** tool.

3. Select objects to sweep or [**MO**de]: *select **both** of the circles and then press <enter> (P1).*

4. Select sweep path or [**A**lignment/**B**ase point/**S**cale/**T**wist]: *select the 2D path (P2).*

5. The **Sweep** command will end automatically and the 3D models will be complete.

6. Use the **Subtract** command to subtract the inner model from the outer model to create a tube. Save the drawing file as **Tube 3D**.

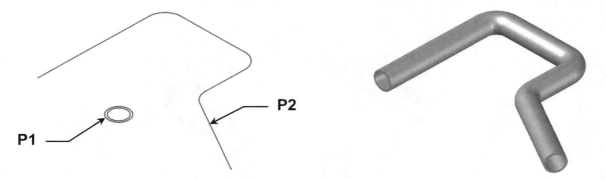

Note: This method used the center of the circles for the base point of the swept path. To use a different base point, refer to the next page.

Sweep Tool....continued

An example of using the Sweep tool with the base point option.

1. Open the sample drawing file **"Pipe - Inch.dwg"** if you are an inch user, or **"Pipe - Metric.dwg"** if you are a metric user.

2. Switch to **Conceptual** style.

3. Select the **Sweep** tool.

4. Select objects to sweep or [**MO**de]: *select <u>both</u> of the circles and then press <enter> (P1).*

5. Select sweep path or [**A**lignment/**B**ase point/**S**cale/**T**wist]: *type in b* (for Base point) *and then press <enter>.*

6. Specify base point: *select the right-hand quadrant on the outer circle using the Quadrant Object Snap (P2).*

7. Select sweep path or [**A**lignment/**B**ase point/**S**cale/**T**wist]: *select the 2D path (P3).*

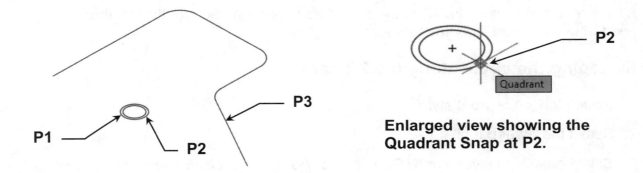

Enlarged view showing the Quadrant Snap at P2.

8. The **Sweep** command will end automatically and the 3D models will be complete.

9. Use the **Subtract** command to subtract the inner model from the outer model to create a tube.

The image below shows the tube has been swept on the outside of the 2D path instead of the center of the 2D path. **Note:** The original 2D path will not be shown on your model.

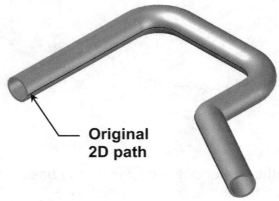

Original 2D path

Sweep Tool....continued

An example of using the Sweep tool with the twist option.

1. Open the sample drawing file **"Bar Twist - Inch.dwg"** if you are an inch user, or **"Bar Twist - Metric.dwg"** if you are a metric user.

2. Switch to **Conceptual** style.

3. Select the **Sweep** tool.

4. Select objects to sweep or [MOde]: *select the square and then press <enter> (P1).*

5. Select sweep path or [**A**lignment/**B**ase point/**S**cale/**T**wist]: *type in t* (for Twist) *and then press <enter>.*

6. Enter twist angle or allow banking for a non-planar sweep path [**B**ank/**E**xpression] <0.000>: *type in 360 and then press <enter>.*

7. Select sweep path or [**A**lignment/**B**ase point/**S**cale/**T**wist]: *select the 2D path (P2).*

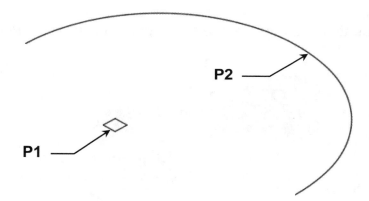

8. The **Sweep** command will end automatically and the 3D model will be complete.

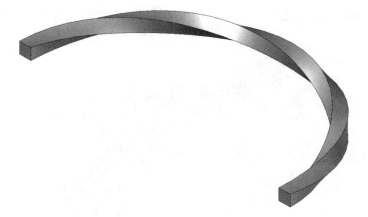

Save the drawing file as **Bar Twist 3D**.

Exercise 7A

1. Open the sample drawing file **"Support Link - Inch.dwg"** if you are an inch user, or **"Support Link - Metric.dwg"** if you are a metric user.

2. Using the **Edit Polyline** tool, join all the lines and arcs to create a closed shape.

3. Select the **SE Isometric** view.

4. Switch to **Conceptual** style.

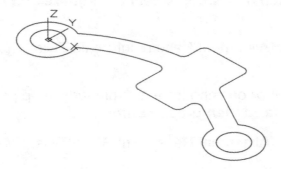

5. Using the **Extrude** tool, extrude the closed shape and 2 inner circles to a height of 0.590" [15 mm].

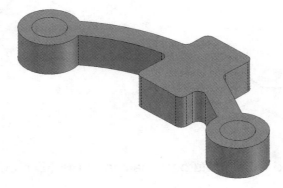

6. Using the **Subtract** tool, subtract the 2 inner circles from the main 3D model.

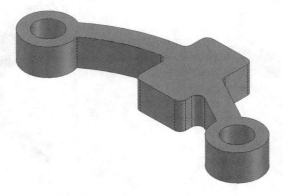

7. Save the drawing file as **Ex-7A**.

Exercise 7B

1. Open the sample drawing file **"Champagne Flute - Inch.dwg"** if you are an inch user, or **"Champagne Flute - Metric.dwg"** if you are a metric user.

2. Using the **Edit Polyline** tool, join all the lines and arcs to create a closed shape.

3. Select the **SE Isometric** view.

4. Switch to **Conceptual** style.

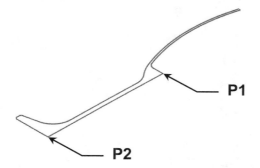

5. Using the **Revolve** tool, revolve the closed shape 360 degrees using **P1** as the axis start point and **P2** as the axis end point.

6. Using the **3D Rotate** tool, rotate the 3D model 90 degrees positive around the **X** axis.

You may wish to experiment using transparency.

7. Save the drawing file as **Ex-7B**.

Exercise 7C

1. Open the sample drawing file **"Rolled Sheet - Inch.dwg"** if you are an inch user, or **"Rolled Sheet - Metric.dwg"** if you are a metric user.

2. Switch to **Conceptual** style.

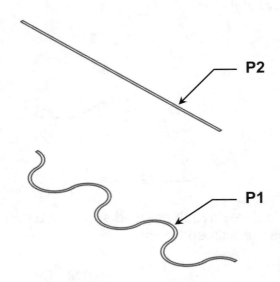

3. Using the **Loft** tool, select the cross sections **P1** and then **P2** to create a lofted 3D model.

4. Save the drawing file as **Ex-7C**.

Exercise 7D

1. Open the sample drawing file **"Handle - Inch.dwg"** if you are an inch user, or **"Handle - Metric.dwg"** if you are a metric user.

2. Switch to **Conceptual** style.

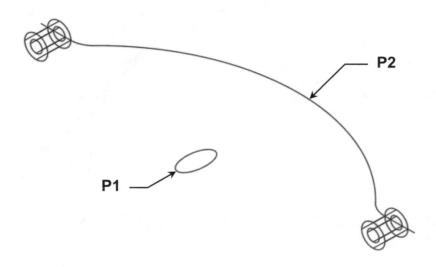

3. Using the **Sweep** tool, select the elliptical shape (**P1**) and sweep it along the 2D path (**P2**) to create a swept 3D model.

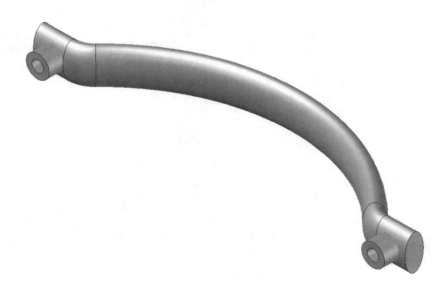

Continued on the next page...

Exercise 7D....continued

4. Using the **Union** tool, union the 2 Cylinders to the swept model.

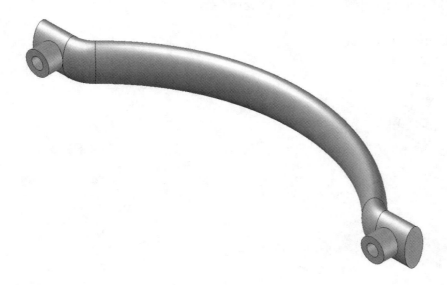

5. Using the **3D Rotate** tool, rotate the 3D model negative 90 degrees (– 90) in the **Y** axis.

4. Save the drawing file as **Ex-7D**.

LEARNING OBJECTIVES

After completing this lesson you will be able to:

1. Shell a 3D Solid Model.
2. Create a Helix.

LESSON 8

Shell Tool

The **Shell** tool allows you to hollow out the insides of a 3D solid model leaving a thin wall. You first specify the face to be shelled and the wall thickness. Some examples of using the Shell tool are shown below.

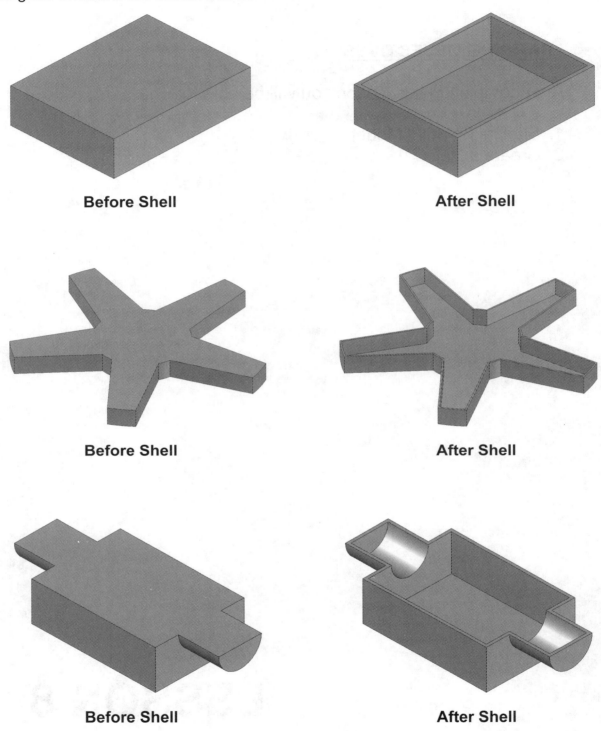

| **Before Shell** | **After Shell** |

| **Before Shell** | **After Shell** |

| **Before Shell** | **After Shell** |

As you can see from the examples, complex 3D solid models can be created using the **Shell** tool.

Shell Tool....continued

There are various methods you can use to initiate the **Shell** tool.

Method 1 – Solid Editing Panel on the 3D Tools Tab of the Drafting and Annotation Workspace.

1. Select the **Shell** tool.

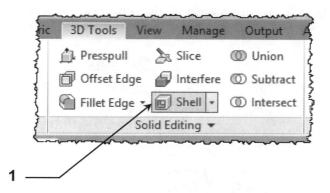

1

Method 2 – Separate drop-down list on the Solid Editing Panel of the Solid Tab of the 3D Modeling Workspace.

1. Select the **Separate** drop-down list.

2. Select the **Shell** tool.

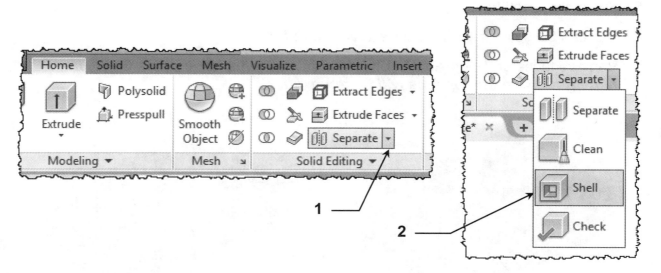

Method 3 – Keyboard entry.

1. On the **Command Line** or in the **Dynamic Input box**, type in *shell* on your keyboard then press *<enter>*.

Shell Tool….continued

An example of using the Shell tool.

1. Start a new drawing file by selecting either **acad.dwt** for inch users, or **acadiso.dwt** for metric users.

2. Select the **SE Isometric** view.

3. Switch to **Conceptual** style.

4. Create a 3D solid Box with the following sizes. You can place the Box anywhere on the screen. It doesn't have to be on 0,0,0.

<u>**Inch users**</u>

Length = 4.000"
Width = 3.000"
Height = 1.000"

<u>**Metric users**</u>

Length = 101.6 mm
Width = 76.2 mm
Height = 25.4 mm

5. Select the **Shell** tool.

6. Select a 3D solid: *select the Box. It will highlight.*

7. Remove faces or [Undo/Add/ALL]: *left click on the top face of the Box. The highlighting on the top face will be removed.*

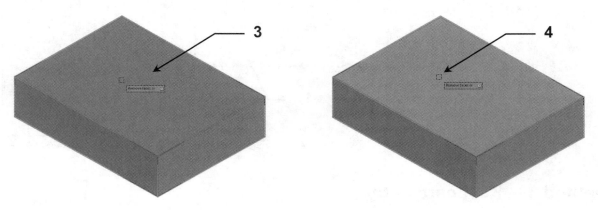

8. Remove faces or [Undo/Add/ALL]: *press <enter>.*

Continued on the next page...

Shell Tool....continued

9. Enter the shell offset distance: *type in 0.1 [2.54] and then press <enter>.*

10. Enter a body editing option
    [**I**mprint/se**P**arate solids/**S**hell/c**L**ean/**C**heck/**U**ndo/e**X**it] <eXit>: *press <enter>.*

11. Enter a solids editing option [**F**ace/**E**dge/**B**ody/**U**ndo/e**X**it] <eXit>: *press <enter> again.*

12. The shelled wall will be complete and the **Shell** command will end automatically.

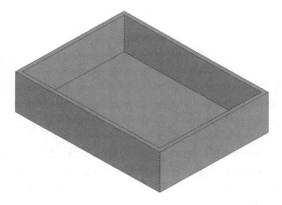

You can shell a 3D solid object using any face, or you can select more than one face. The example below shows the Box with the top and end face selected and with the completed model after adding the shell offset distance.

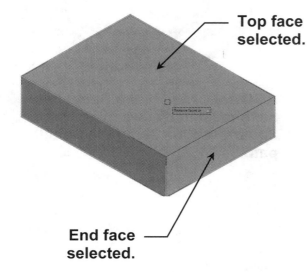

Top face selected.

End face selected.

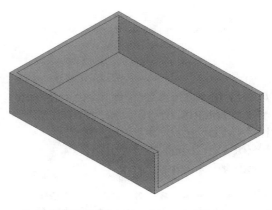

The completed model after adding the shell offset distance.

Helix Tool

The **Helix** tool allows you to construct a wireframe Helix or Spiral, which can then be turned into a 3D solid model by sweeping an object along its path, such as a Spring, Coil or Screw Thread. When creating the Helix you can specify the base and top radius, the height, number of turns, turn height and twist direction. The default twist direction is counter-clockwise. Some examples of using the Helix tool are shown below.

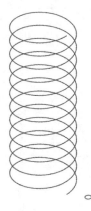

A Helix with a 10 mm base and top radius, a turn height of 5 mm, 12 turns, and a total height of 60 mm in a counter-clockwise direction.

There is also a 2 mm diameter Circle that needs to be swept along the Helix path.

A 3D solid model of a Spring has been created by sweeping the 2 mm diameter Circle along the Helix path.

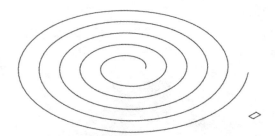

A Spiral with a 5 mm base radius, a top radius of 35 mm, a turn height of zero, 5 turns, and a height of zero in a counter-clockwise direction.

There is also a Square section that needs to be swept along the Spiral path.

A 3D solid model of a Spiral has been created by sweeping the Square section along the Spiral path.

As you can see from the examples, complex 3D solid models can be created using the **Helix** tool.

Helix Tool....continued

There are various methods you can use to initiate the **Helix** tool.

Method 1 – Draw drop-down list of the Home Tab on the Drafting and Annotation Workspace.

1. Select the Draw Panel drop-down list.
2. Select the **Helix** tool.

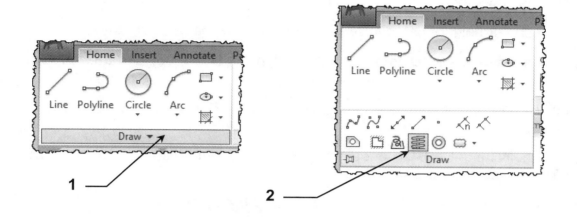

Method 2 – Draw drop-down list of the Home Tab on the 3D Modeling Workspace.

1. Select the Draw Panel drop-down list.
2. Select the **Helix** tool.

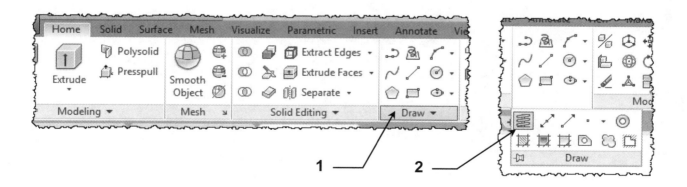

Method 3 – Keyboard entry.

1. On the **Command Line** or in the **Dynamic Input box**, type in *helix* on your keyboard then press *<enter>*.

Helix Tool....continued

An example of using the Helix tool.

1. Start a new drawing file by selecting either **acad.dwt** for inch users, or **acadiso.dwt** for metric users.

2. Select the **SE Isometric** view.

3. Switch to **Conceptual** style.

4. Select the **Helix** tool.

5. Specify center point of base: *left click anywhere in the drawing area.*

6. Specify base radius or [Diameter] <1.000>: *type in 0.394 (10) and then press <enter>.*

7. Specify top radius or [Diameter] <0.394>: *type in 0.394 (10) and then press <enter>.*

8. Specify helix height or [Axis endpoint/Turns/turn Height/tWist] <1.000>: *type in t* (for Turns) *and then press <enter>.*

9. Enter number of turns <3.000>: *type in 12 and then press <enter>.*

10. Move your mouse in an upward direction (Z+).

11. Specify helix height or [Axis endpoint/Turns/turn Height/tWist] <1.000>: *type in 2.362 [60] and then press <enter>.*

12. The Helix will be complete and the **Helix** command will end automatically.

13. Save this drawing file as **Spring 3D** but keep the file open.

Continued on the next page...

Helix Tool....continued

14. Create a Circle with a radius of **0.039" [1 mm]** and place it approximately in the location shown. The Circle position is not important as it will be used to sweep along the Helix path.

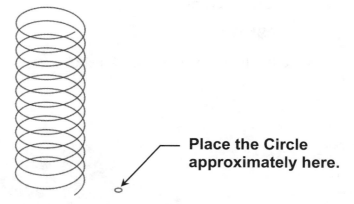

Place the Circle approximately here.

15. Select the **Sweep** tool.

16. Select objects to sweep or [Mode]: *select the Circle and then press <enter>.*

17. Select sweep path or [**A**lignment/**B**ase point/**S**cale/**T**wist]: *type in b* (for Base point) *and then press <enter>.*

18. Specify base point: *left click on the quadrant shown below using the Quadrant Snap and then press <enter>.*

Select this quadrant.

Quadrant

19. Select sweep path or [**A**lignment/**B**ase point/**S**cale/**T**wist]: *select the Helix.*

20. The model will be complete and the **Sweep** command will end automatically.

Note: Depending on your system, it might take a few seconds to compute the sweep path. So don't be alarmed and think that your system has crashed.

21. Resave the drawing file as **Spring 3D**

Exercise 8A

1. Open the sample drawing file **"Cover - Inch.dwg"** if you are an inch user, or **"Cover - Metric.dwg"** if you are a metric user.

2. Switch to **Conceptual** style.

3. Select the **Shell** tool.

4. Shell the top face and end face as shown, with a shell offset distance of 0.050 [1.27].

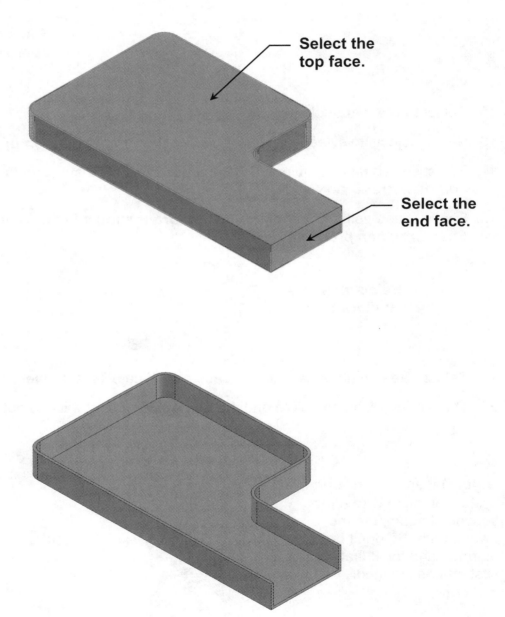

Select the top face.

Select the end face.

5. Save the drawing file as **Ex-8A**

Exercise 8B

1. Start a new drawing file by selecting either **acad.dwt** for inch users, or **acadiso.dwt** for metric users.

2. Select the **SE Isometric** view.

3. Switch to **Conceptual** style.

4. Select the **Helix** tool and create a Helix with the following sizes.

<u>**Inch users**</u>

Base radius = 0.787"
Top radius = 0.394"
Turns = 8
Turn height = 0.197"

<u>**Metric users**</u>

Base radius = 20 mm
Top radius = 10 mm
Turns = 8
Turn height = 5 mm

5. Create a Circle with a radius of 0.059 [1.5] and place it near the Helix.

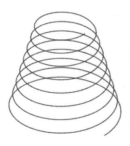

6. Select the **Sweep** tool and sweep the Circle along the Helix path.

7. Save the drawing file as **Ex-8B**

Exercise 8C

1. Start a new drawing file by selecting either **acad.dwt** for inch users, or **acadiso.dwt** for metric users.

2. Select the **SE Isometric** view.

3. Switch to **Conceptual** style.

4. Select the **Helix** tool and create a Spiral with the following sizes.

<u>Inch users</u>	<u>Metric users</u>
Base radius = 0.250"	Base radius = 6.35 mm
Top radius = 1"	Top radius = 25.4 mm
Turns = 6	Turns = 6
Turn height = 0	Turn height = 0

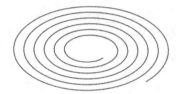

5. Create a Circle with a radius of 0.031 [0.79] and place it near the Spiral.

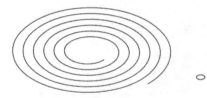

6. Select the **Sweep** tool and sweep the Circle along the Spiral path.

7. Save the drawing file as **Ex-8C**

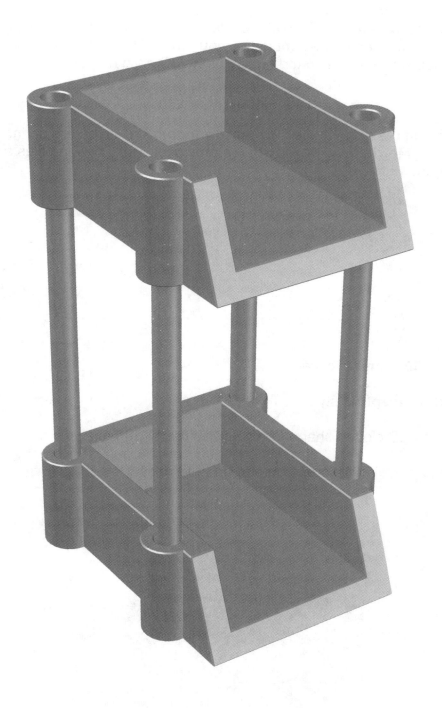

PROJECT 1
Stackable Junk Tray

Project 1: Stackable Junk Tray

In this project you will create 3D solid models of 2 junk trays and 4 spacing legs. These 3D solid models will then be assembled so you have a 2-tier stackable junk tray system. You will also learn how to refine the quality of the 3D solid models to make them more realistic.

1. Open the sample drawing file **"Project 1 - Inch.dwg"** if you are an inch user, or **"Project 1 - Metric.dwg"** if you are a metric user. Both drawing files are located in the **"Project 1"** folder.

2. Select the **SE Isometric** view.

3. Switch to **Conceptual** style.

4. **Extrude** the 2D closed shape to a height of 3.150" [80 mm].

5. **Rotate** the model 90 degrees positive in the X axis.

6. Save the drawing file as **Stackable Junk Tray** but keep the drawing file open.

Project 1: Stackable Junk Tray….continued

7. Select the **Shell** tool and shell the top face and the end angled face with a shell offset distance of 0.394" [10 mm].

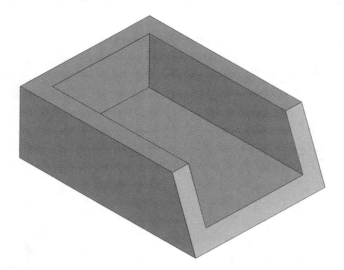

8. Create a Cylinder with a radius of 0.394" [10 mm] and place it at the bottom left-hand corner (**P1**) as shown using the Endpoint Snap. Place the other end of the Cylinder at the top left-hand corner (**P2**) as shown using the Endpoint Snap.

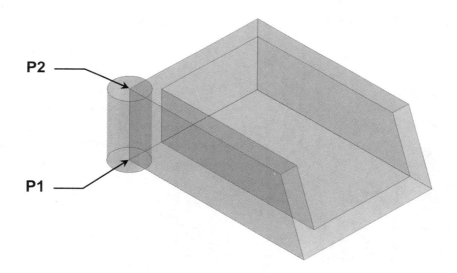

Note: I have changed both models to transparency for clarity.

9. Resave your drawing file but keep it open.

Project 1: Stackable Junk Tray....continued

10. Select the **Move** tool and move the Cylinder to the corner of the larger model using the **Quadrant Snap** on the Cylinder as the base point, and then by using the **Endpoint Snap** to snap to the corner of the larger model, as show below.

Use this Quadrant as the base point.

Snap to the corner using the Endpoint Snap.

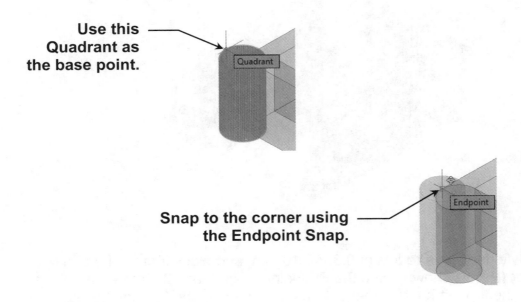

11. Select the **Copy** tool and copy the Cylinder to the opposite back corner of the larger model using the **Endpoint Snap** for the base point and copied location.

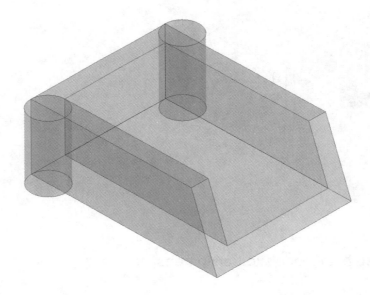

12. Resave your drawing file but keep it open.

Project 1: Stackable Junk Tray....continued

13. Turn on **Ortho Mode** (*F8*).

14. Select the **Copy** tool.

15. Select objects: ***select both the Cylinders and then press <enter>.***

16. Specify base point or [**D**isplacement/m**O**de] <Displacement>: ***select the top center point of the left-hand Cylinder.***

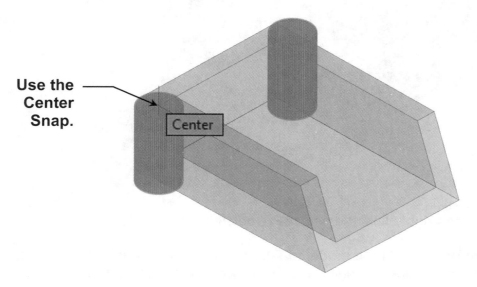

Use the Center Snap. → Center

17. Specify second point or [**A**rray] <use first point as displacement>: ***move your mouse along the X+ axis and then type in 2.756 [70] and then press <enter>.***

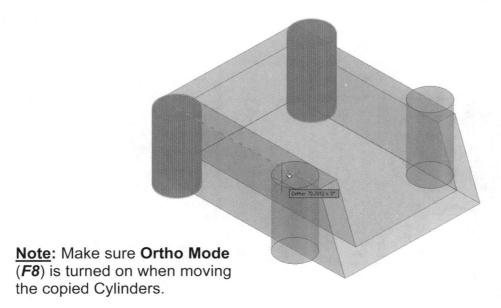

Ortho: 70.2912 < 0°

Note: Make sure **Ortho Mode** (*F8*) is turned on when moving the copied Cylinders.

18. Specify second point or [**A**rray/**E**xit/**U**ndo] <Exit>: ***press <enter> again to end the Copy command.***

Project 1: Stackable Junk Tray....continued

19. Select the **Union** tool and union the 4 Cylinders to the larger model.

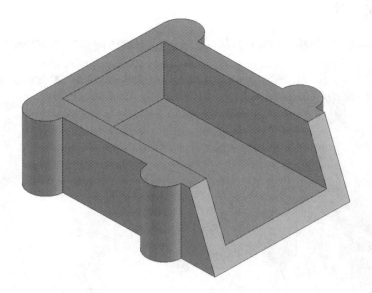

20. Select the model then right click and select **Properties** from the list.

21. Change the **Transparency** of the model to **70**. This will make it easier for the next step. Press the *Esc* key after changing the Transparency to deselect the model.

22. Create 2 more Cylinders with a radius of 0.197" [5 mm] and a height of 0.394" [10 mm]. Place them at the top and bottom of the front right-hand lug as shown.

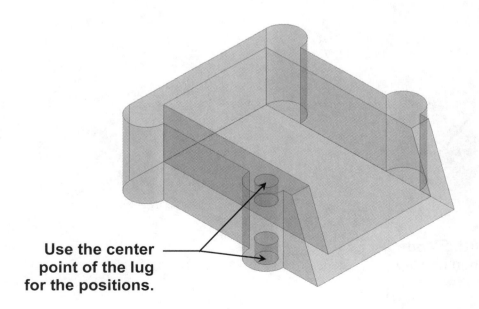

Use the center point of the lug for the positions.

23. Resave your drawing file but keep it open.

Project 1: Stackable Junk Tray....continued

24. Turn off **Ortho Mode** (*F8*). This will make it easier for the next step.

25. Select the **Copy** tool.

26. Select objects: *using a Window Selection starting at P1 and ending at P2, select the 2 small Cylinders and then press <enter>.*

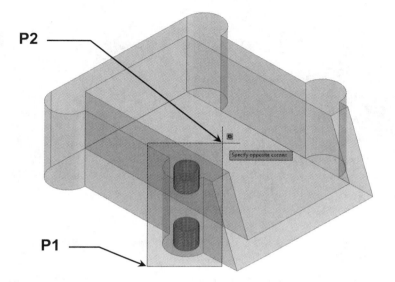

27. Specify base point or [**D**isplacement/m**O**de] <Displacement>: *select the top center point of the top Cylinder.*

28. Specify second point or [**A**rray] <use first point as displacement>: *copy the 2 Cylinders to the front left-hand lug using the top center point of the lug for the location and then left click to place the 2 Cylinders.*

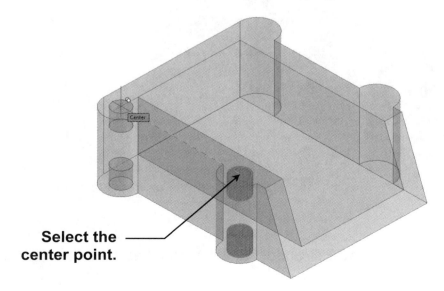

Select the center point.

Project 1: Stackable Junk Tray....continued

29. Specify second point or [**A**rray/**E**xit/**U**ndo] <Exit>: *copy the 2 Cylinders to the top right-hand lug by using the Center Snap and then left clicking with your mouse.*

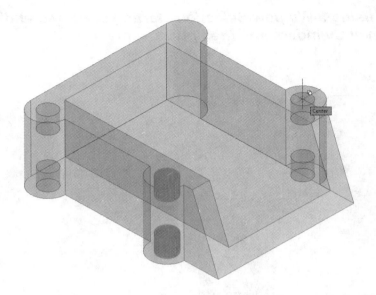

30. Specify second point or [**A**rray/**E**xit/**U**ndo] <Exit>: *copy the 2 Cylinders to the top left-hand lug by using the Center Snap and then left clicking with your mouse.*

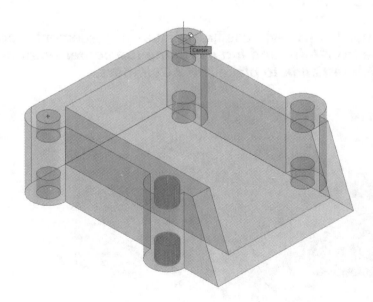

31. Specify second point or [**A**rray/**E**xit/**U**ndo] <Exit>: *press <enter> to end the Copy command.*

32. Resave your drawing file but keep it open.

Project 1: Stackable Junk Tray....continued

33. Select the **Subtract** tool.

34. Select objects: *left click on the main model and then press <enter>.*

35. Select objects: *using a Window Selection starting at P1 and ending at P2, select the 8 small Cylinders and then press <enter>.*

P1

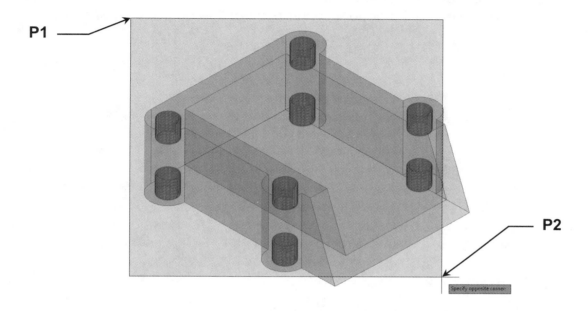

P2

36. Select the model and then right click and select **Properties** from the list, and then change the **Transparency** to **0** (zero).

Your completed 3D solid model of a Stackable Junk Tray should look like the image below. The next stage is to create 4 spacing legs and a copy of the Junk Tray and place them into position.

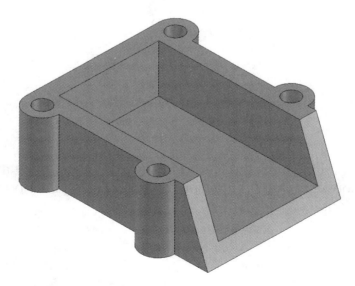

37. Resave your drawing file but keep it open.

Project 1: Stackable Junk Tray....continued

38. Create a solid Cylinder with a radius of 0.197" [5 mm] and a height of 5.118" [130 mm] and place it anywhere in the drawing area.

39. Select the main model and the Cylinder then right click and select **Properties** from the list. Change the **Transparency** to **70**.

40. Select the **Move** tool and select the Cylinder. Use the center point at the bottom of the Cylinder and move it to the bottom center point of the top front left-hand hole.

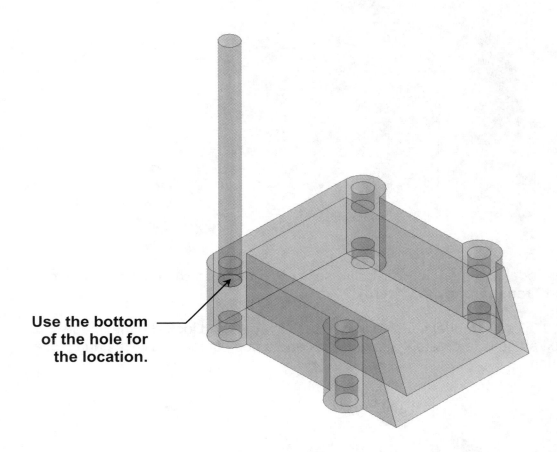

Use the bottom of the hole for the location.

41. Select the **Copy** tool and select the Cylinder. Use the center point at the bottom of the Cylinder and copy it to the 3 remaining top holes, using their bottom center points for the locations.

Continued on the next page...

Project 1: Stackable Junk Tray....continued

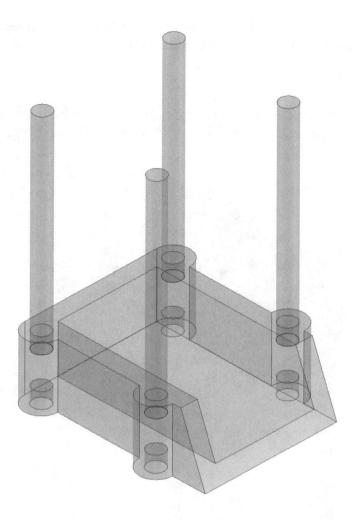

42. Select the **Copy** tool then select the main model (not the 4 Cylinders). Using the center point at the top of the bottom front left-hand hole for the base point, copy the model up to the top of the front left-hand Cylinder using the center point for the location, as shown below.

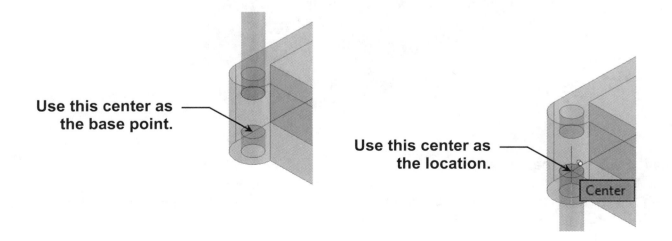

Use this center as the base point.

Use this center as the location.

Center

Project 1: Stackable Junk Tray....continued

43. Select the main model and the 4 Cylinders then right click and select **Properties** from the list. Change the **Transparency** to **0** (zero). Press the *Esc* key to deselect the models.

Your completed 3D solid model of 2 Stackable Junk Trays with their spacing legs should look like the image below. The next stage is to make the completed models look more realistic.

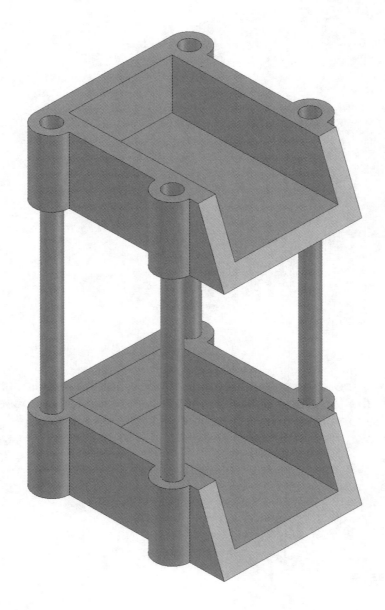

44. Switch to **Realistic** style.

45. Select both of the junk trays and then right click and choose **Properties** from the list. (Do not select the 4 spacing legs.)

Project 1: Stackable Junk Tray....continued

46. In the **Properties** dialog box, select the **Color** drop-down list and then click on **Select Color**.

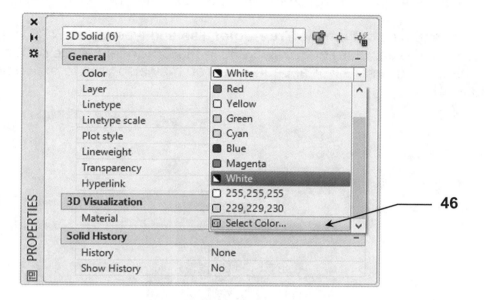

46

47. In the **Select Color** dialog box, select a color for the 2 junk trays. I have chosen yellow, but you may choose another color if you wish.

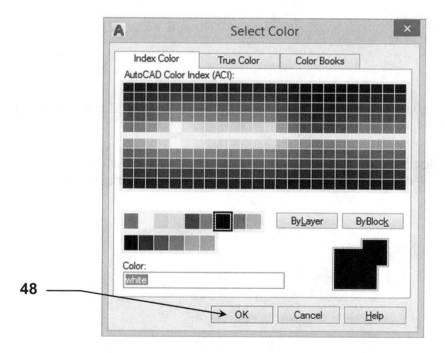

48

48. Click on **OK** when you have chosen your color.

Note: You can also experiment with different colors by selecting the **True Color** or **Color Books** Tabs.

Project 1: Stackable Junk Tray....continued

49. Press the *Esc* key on your keyboard to deselect the 2 junk trays.

50. Select the 4 spacing legs then repeat **Steps 45-49** to change the color of the legs.

I have chosen a color from the **True Color** Tab and typed in an RGB number (RGB means Red, Green, Blue) of 229,229,230 in the **Color** field at the bottom of the dialog box. This color gives a silver appearance. You may also wish to experiment with the color slide bar.

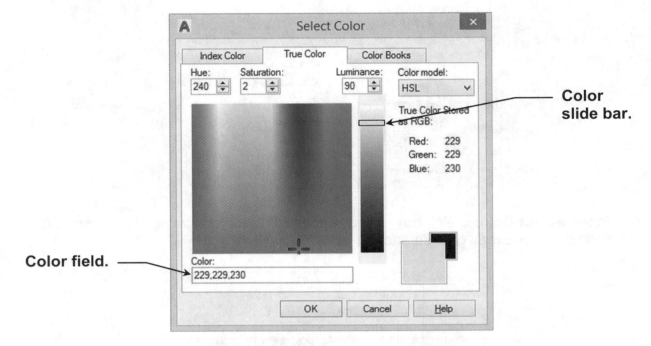

Color field.

Color slide bar.

51. Resave your drawing file as **Stackable Junk Tray**, overwriting the existing drawing file.

If you select the **Color Books** Tab you will find a wide range of interesting color books like the **PANTONE+ Premium Metallics Coated** book. It is worth experimenting with various color books to get a more realistic model.

Project 1: Stackable Junk Tray....continued

You may also wish to experiment with different **Visual Styles** or changing an existing style by tweaking various properties. This is achieved by selecting the Visual Styles Manager dialog box.

Selecting the Visual Styles Manager dialog box.

1. Left click on the word **"Realistic"** at the top left-hand corner of the drawing area. Yours may say **"Conceptual"** or **"2D Wireframe"**.

2. Select **Visual Styles Manager** from the list.

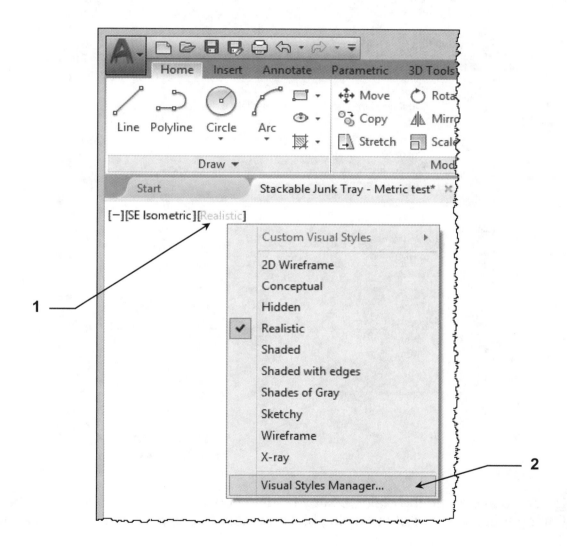

3. The **Visual Styles Manager** dialog box will open.

Continued on the next page...

Project 1-15

Project 1: Stackable Junk Tray....continued

Current Visual Style.

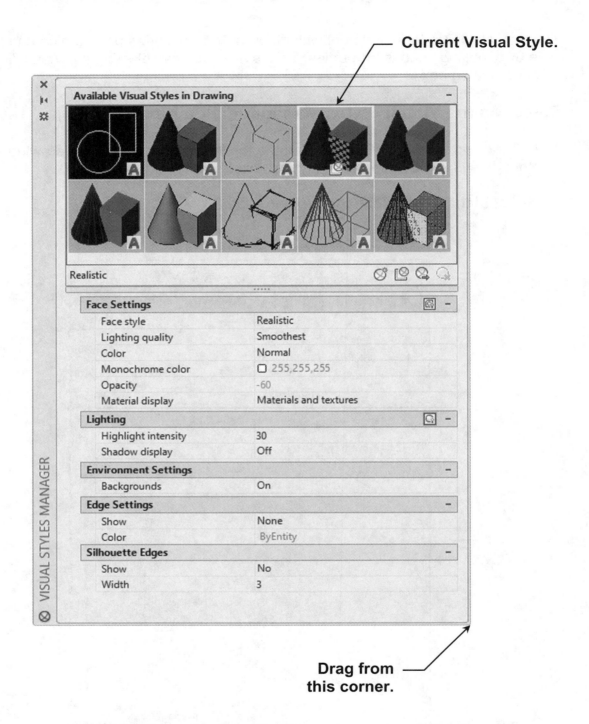

Drag from this corner.

Note: You can expand the dialog box by holding down the left mouse button at the bottom right-hand corner of the dialog box and then dragging the mouse to expand.

The current Visual Style is **Realistic** and highlighted in the 10 images shown in the dialog box. If you hover your mouse over any of the images it will tell you what styles they are.

Continued on the next page...

Project 1: Stackable Junk Tray....continued

Experiment with changing various fields in the **Visual Styles Manager** dialog box. For example, change the **Lighting - Highlight Intensity** field to **70**. This will provide a more intense light to the model.

Change to 70

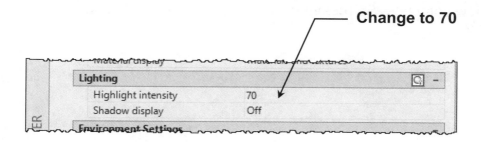

Close down the **Visual Styles Manager** dialog box when you have finished experimenting with the various fields.

Try changing the Visual Style of the completed model to **Sketchy**. This gives the appearance of a hand-drawn sketch.

Visual Style changed to Sketchy.

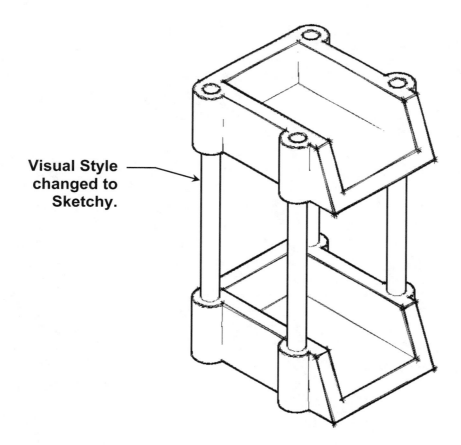

Note: If you were to save the drawing file with the Sketchy Visual Style and close the drawing file, you could always change the Visual Style after re-opening the drawing file again.

Project 1: Stackable Junk Tray....continued

There are many ways you could have created the Junk Tray and this was just one way of achieving it. As you get more experienced in 3D modeling you may find easier or different ways of creating 3D models.

Project 1 is now complete. So resave the drawing file and close it down, then progress on to **Project 2**.

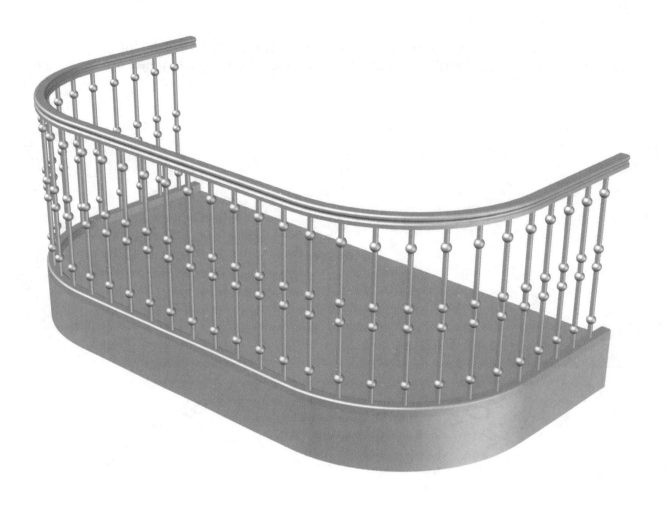

PROJECT 2
Ornate Balcony

Project 2: Ornate Balcony

In this project you will create 3D solid models for an Ornate Balcony. These will include a base, bottom and top rails with their profiles, and the spindles. You will also learn how to refine the quality of the 3D solid models to make them more realistic.

1. Open the sample drawing file **"Project 2 - Inch.dwg"** if you are an inch user, or **"Project 2 - Metric.dwg"** if you are a metric user. Both drawing files are located in the **"Project 2"** folder.

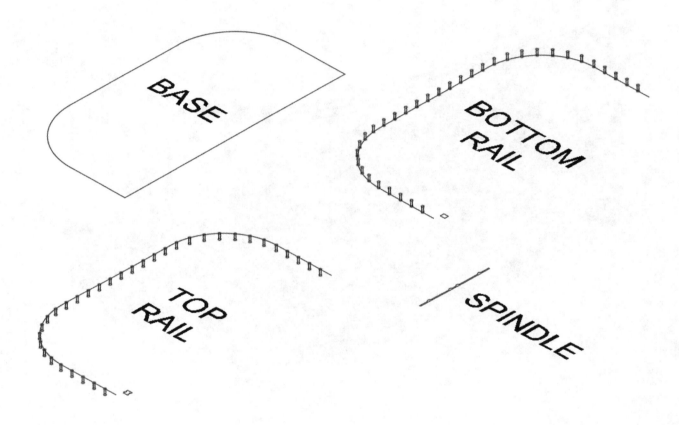

You will notice that the parts are labeled for easy identification. The labels can be removed at any time you choose by selecting a label and pressing the **Delete** key. You will also notice a series of Cylinders that are aligned with the bottom and top rail paths. These Cylinders will become the holes in both rails after the rail profiles have been swept along their paths and then subtracted.

2. Switch to **Conceptual** style.

3. **Extrude** the base to a height of 12" [304.8 mm].

4. Save the drawing file as **Ornate Balcony** but keep the drawing file open.

Project 2: Ornate Balcony....continued

You may have noticed that when in **Conceptual** style there are heavy black lines around the edges of a 3D solid model. These are called **"Silhouette Edges"**. To remove these heavy lines select the **Visual Styles Manager.** Under **Silhouette Edges** click on the **Show** field and change it to **No**.

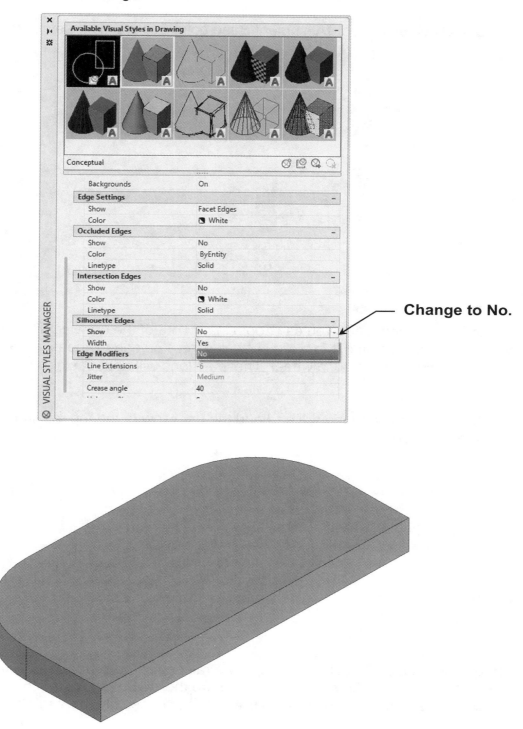

Change to No.

The balcony base is now complete and the next stage is to create the top and bottom rails by sweeping their profiles along the paths.

Project 2: Ornate Balcony....continued

5.   Select the **Sweep** tool.

6.   Select objects to sweep or [**MO**de]: *select the top rail profile and then press <enter> (P1).*

7.   Select sweep path or [**A**lignment/**B**ase point/**S**cale/**T**wist]: *select the path (P2).*

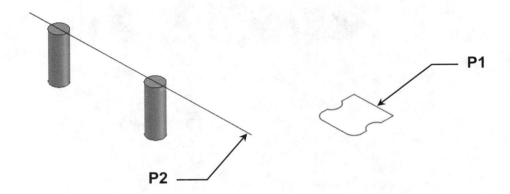

Note: During commands you can zoom in and out of the drawing area with the wheel on your mouse by scrolling it forward or backward. You can also pan around the drawing by holding down the mouse wheel. These actions will not affect the command.

8.   The **Sweep** command will end and the end view of the top rail should look like the image below.

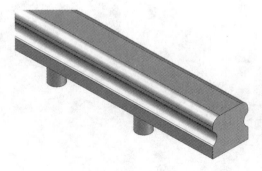

Note: If you get unexpected results or the rail profile looks upside down, use the **Undo** button to go back to the start then make sure the **UCS** is in **World** coordinates by typing in *ucs <enter>*, then typing in *w <enter>*. If that doesn't help try rotating the rail profile 180 degrees in the Z axis.

9.   Resave your drawing file as **Ornate Balcony** but keep the drawing file open.

Project 2: Ornate Balcony....continued

10. Select the **Subtract** tool.

11. Select objects: *select the top rail and then press <enter>.*

12. Select objects: *select all the Cylinders below the top rail using a Window Selection starting at P1 and ending at P2 and then press <enter>.*

<u>Note</u>: Make sure you do not include the rail in the Window selection.

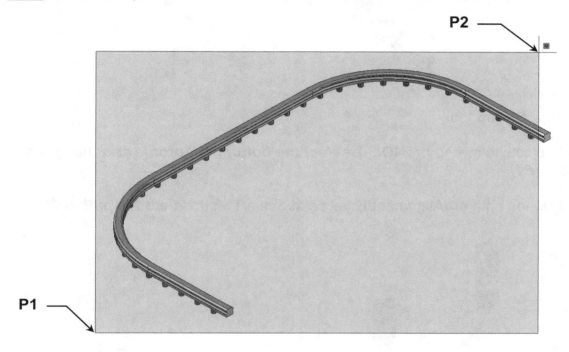

P2

P1

<u>Note</u>: If you prefer you can select all the Cylinders using 2 or even 3 Window Selections.

The top rail is now complete.

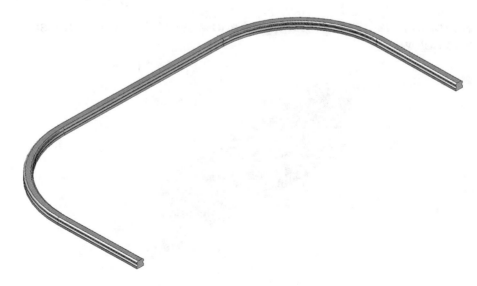

Project 2: Ornate Balcony....continued

If you use the **Orbit** tool and zoom to the underside of the top rail you will see that the Cylinders have now become blind holes in the rail.

13. Select the **Sweep** tool.

14. Select objects to sweep or [**MO**de]: *select the bottom rail profile and then press <enter> (P1).*

15. Select sweep path or [**A**lignment/**B**ase point/**S**cale/**T**wist]: *select the path (P2).*

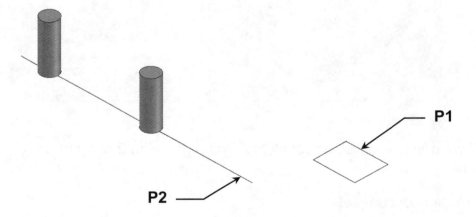

16. The **Sweep** command will end and the end view of the bottom rail should look like the image below.

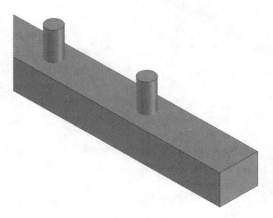

Project 2: Ornate Balcony....continued

17. Select the **Subtract** tool.

18. Select objects: *select the bottom rail and then press <enter>.*

19. Select objects: *select all the Cylinders above the bottom rail using a Window Selection starting at P1 and ending at P2 and then press <enter>.*

<u>Note:</u> Make sure you do not include the rail in the Window selection.

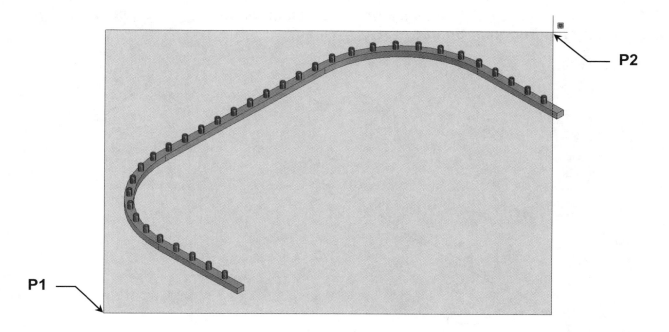

<u>Note:</u> If you prefer you can select all the Cylinders using 2 or even 3 Window Selections.

The bottom rail is now complete.

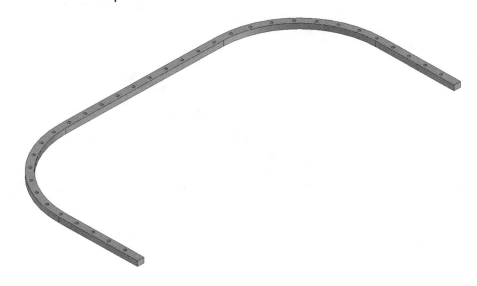

Project 2: Ornate Balcony....continued

20. Resave your drawing file as **Ornate Balcony** but keep the drawing file open.

21. Select the **Revolve** tool.

22. Select objects to revolve or [**MO**de]: *select the spindle and then press <enter>.*

23. Specify axis start point or define axis by [**Object/X/Y/Z**] <Object>: *left click on the endpoint of the long straight line at P1 using the Endpoint Snap.*

24. Specify axis end point: *left click on the endpoint of the long straight line at P2 using the Endpoint Snap.*

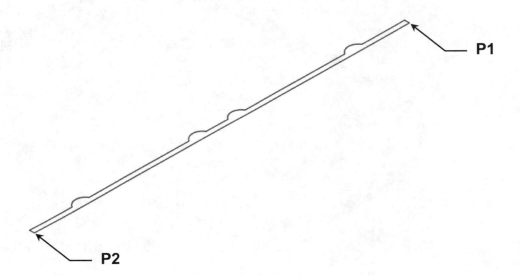

25. Specify angle of revolution or [**ST**art angle/**R**everse/**EX**pression] <360>: *type in 360 and then press <enter>.*

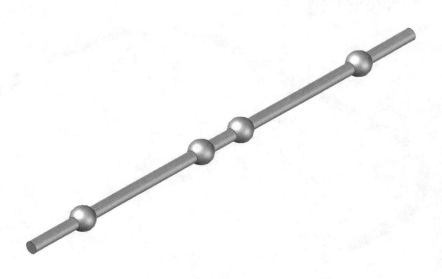

Project 2: Ornate Balcony....continued

26. Select the **3D Rotate** tool and rotate the spindle around the X axis by 90 degrees positive.

The spindle is now complete.

27. Resave your drawing file as **Ornate Balcony** but keep the drawing file open.

The next stage is to move the spindle and locate it into one of the holes on the bottom rail, and then copy that spindle to the remaining 33 holes on the bottom rail.

28. Switch to **2D Wireframe** style. This will make it easier to see the holes in the bottom rail. Alternatively you could select the bottom rail and spindle, right click and select **Properties** from the list. Then change the **Transparency** of the bottom rail and spindle to **70** in the **Properties** dialog box.

29. Turn off **Ortho Mode** (*F8*). This will make it easier when moving the spindle and when copying the spindle to the other holes in the bottom rail.

Project 2: Ornate Balcony….continued

30. Select the **Move** tool.

31. Select objects: *left click on the spindle and then press <enter>.*

32. Specify base point or [Displacement] <Displacement>: *left click on the center point at the bottom of the spindle at P1 using the Center Snap.*

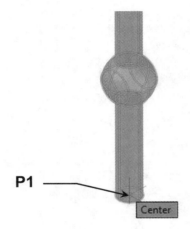

33. Specify second point or <use first point as displacement>: *move the spindle over to the back right-hand hole on the bottom rail and place it on the center point at the bottom of the hole at P2 using the Center Snap.*

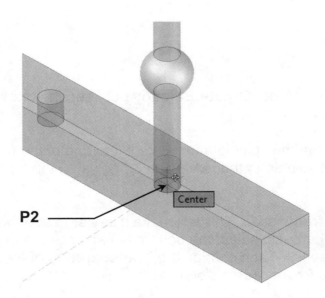

34. The **Move** tool will end automatically and the spindle will be in position.

Project 2: Ornate Balcony....continued

35. Select the **Copy** tool.

36. Select objects: *left click on the spindle and then press <enter>.*

37. Specify base point or [**D**isplacement/m**O**de] <Displacement>: *left click on the center point at the bottom of the spindle at P1 using the Center Snap.*

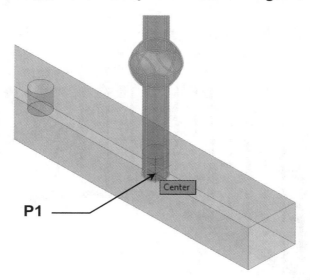

38. Specify second point or [**A**rray] <use first point as displacement>: *move the copied spindle to the next hole in the line and left click on the center point at the bottom of the hole at P2 using the Center Snap.*

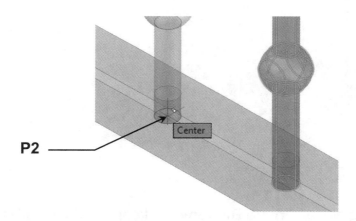

Note: During the **Copy** command you can zoom in and out of the drawing area. You can also pan around. It will not affect the command.

39. Specify second point or [**A**rray/**E**xit/**U**ndo] <Exit>: *move to the next hole in the line and repeat the process for the rest of the holes on the bottom rail.*

40. After placing the copied spindle in the last hole press *<enter>* to end the **Copy** command.

Project 2: Ornate Balcony....continued

41. Switch to **Conceptual** style or if you changed the **Transparency** of the bottom rail and spindles, change it back to zero.

Your model should look like the image below.

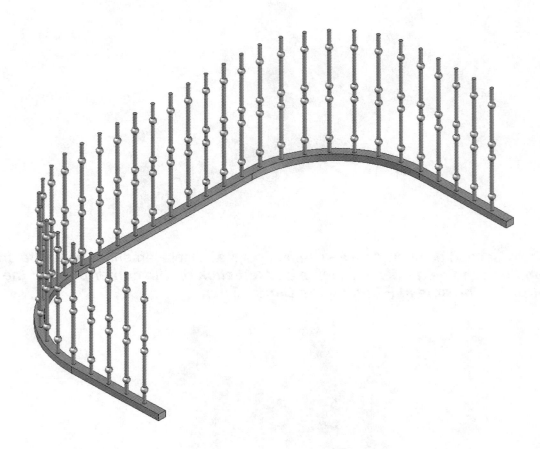

42. Resave your drawing file as **Ornate Balcony** but keep the drawing file open.

The next stage is to place the top rail onto the top of the spindles using the **Move** tool.

43. Select the top rail and change its **Transparency** to **70**.

Project 2: Ornate Balcony....continued

44. Select the **Move** tool.

45. Select objects: *left click on the top rail and then press <enter>.*

46. Specify base point or [**D**isplacement] <Displacement>: *left click on the center point at the top of the front right-hand hole at P1 using the Center Snap.*

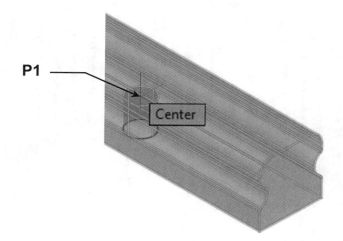

47. Specify second point or <use first point as displacement>: *move the top rail over to the front right-hand spindle and place it on the center point at the top of the spindle at P2 using the Center Snap.*

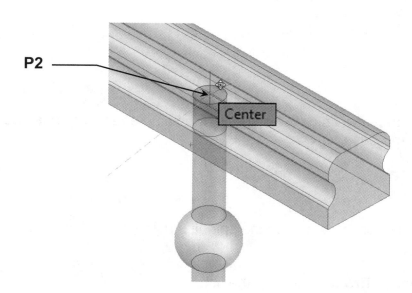

48. The **Move** tool will end automatically and the top rail will be placed into position.

49. Change the **Transparency** of the top rail back to zero.

Project 2: Ornate Balcony....continued

Your model should look like the image below.

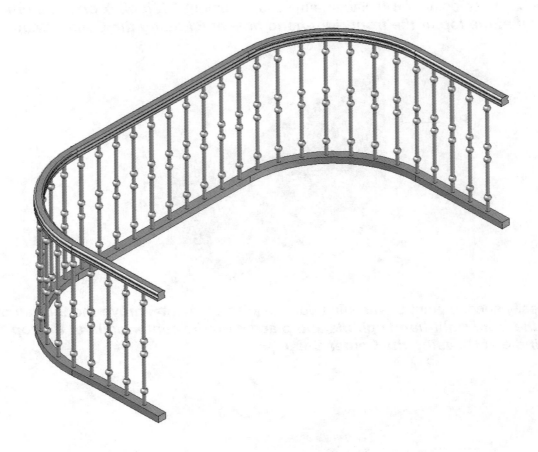

The next stage is to change the color of both rails, the spindles and the base before positioning them on to the base.

50. Switch to **Realistic** style.

51. Select the top and bottom rails and all the spindles using a Window Selection. Do not select the base.

52. Right click and choose **Properties** from the list.

Project 2: Ornate Balcony....continued

53. In the **Properties** dialog box, select the **Color** drop-down list and then click on **Select Color**.

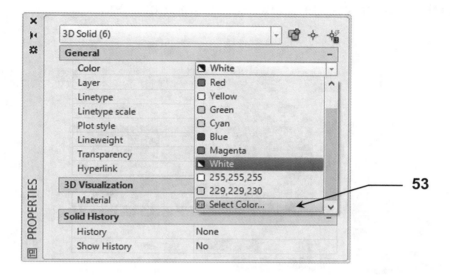

54. In the **Color** field, type in **41** (this color gives a gold appearance).

55. Select **OK** when you have entered the color number.

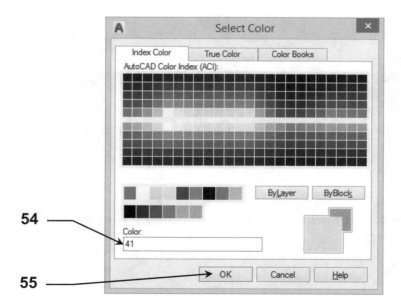

56. Press the *Escape* key on your keyboard to deselect the rails and spindles.

57. Select the base.

Project 2: Ornate Balcony....continued

58. Repeat **Steps 52-55** but enter the number **229,229,230** in the **Color** field. This gives it a light grey color.

59. Select **OK** when you have entered the color number.

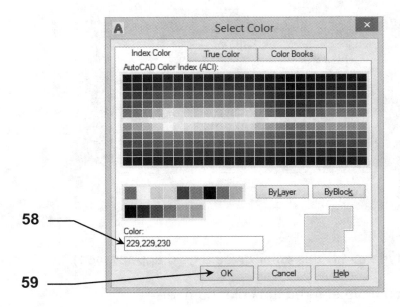

60. Press the **_Escape_** key on your keyboard to deselect the base.

The final stage is to position both rails and all the spindles on to the base.

61. Select the **Move** tool.

62. Select objects: **_select both rails and all the spindles using a Window Selection and then press <enter>._**

Project 2: Ornate Balcony....continued

63. Specify base point or [Displacement] <Displacement>: *left click on the bottom front corner edge of the bottom rail using the Endpoint Snap at P1.*

P1 — Endpoint

64. Specify second point or <use first point as displacement>: *move both rails and all the spindles over to the base and left click on the top front corner edge of the base using the Endpoint Snap at P2.*

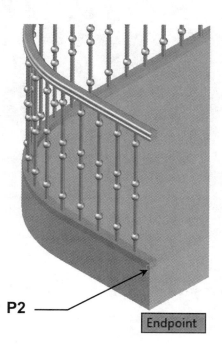

P2 — Endpoint

65. The **Move** tool will end automatically and the rails and spindles will be placed into position.

Project 2: Ornate Balcony….continued

66. Change the drawing view to **SW Isometric**.

The completed **Ornate Balcony** is now complete and should look like the image below.

Project 2 is now complete. So resave the drawing file and close it down, then progress on to **Project 3**.

Project 3: Working Platform

In this project you will create 3D solid models for a Working Platform. All of the objects are created from 3D solid Boxes and Cylinders.

1. Start a new drawing file by selecting either **acad.dwt** for inch users, or **acadiso.dwt** for metric users.

The image below is the main working platform without the steps and platform floor, showing the labeled parts and the quantity of each part that is required.

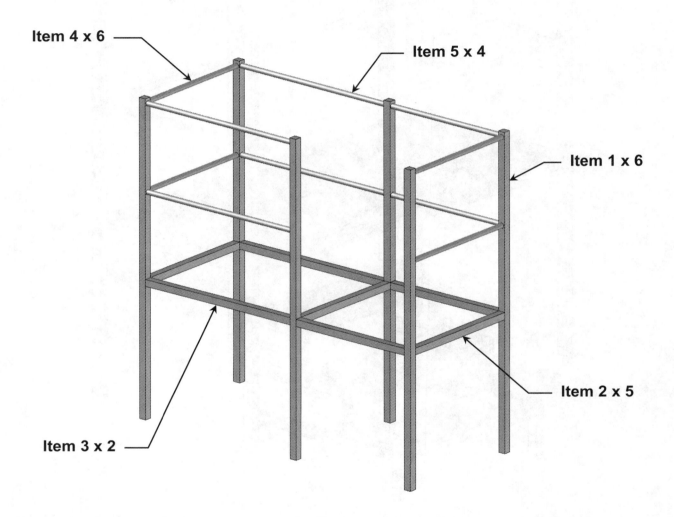

It is entirely up to you which way you choose to create the solid Boxes and Cylinders. I prefer to create them all as uprights and then rotate the cross members and cylindrical handrails afterward.

It is easier to create one of each item and then simply copy them. The sizes of the solid Boxes and Cylinders are given on the next few pages, as well as images showing the positions of the cross members and cylindrical handrails.

Start by creating the main platform framework and assembling it, then create the steps.

Project 3: Working Platform....continued

Sizes for the solid Boxes.

Item 1

Inch users

Length = 1.575"
Width = 1.575"
Height = 71.654"

Metric users

Length = 40 mm
Width = 40 mm
Height = 1820 mm

Item 2

Inch users

Length = 1.575"
Width = 1.575"
Height = 28.346"

Metric users

Length = 40 mm
Width = 40 mm
Height = 720 mm

Item 3

Inch users

Length = 1.575"
Width = 1.575"
Height = 37.795"

Metric users

Length = 40 mm
Width = 40 mm
Height = 960 mm

Sizes for the solid Cylinders.

Item 4

Inch users

Radius = 0.590"
Height = 28.346"

Metric users

Radius = 15 mm
Height = 720 mm

Item 5

Inch users

Radius = 0.590"
Height = 37.795"

Metric users

Radius = 15 mm
Height = 960 mm

Project 3: Working Platform….continued

2. Create **Item 1** and place it anywhere in the drawing area.

3. Create **Item 2** and rotate it 90 degrees around the X axis.

4. Move **Item 2** to the bottom corner of **Item 1**.

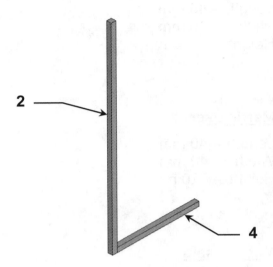

5. Copy **Item 1** and move it to the other end of **Item 2**.

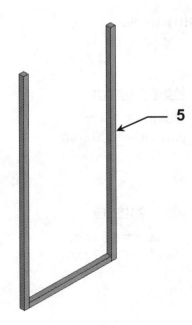

6. Save the drawing file as **Working Platform** but keep the drawing file open.

Project 3: Working Platform….continued

7. Copy **Item 2** and place it anywhere in the drawing area.

8. Rotate the copied **Item 2** around the Z axis by 90 degrees.

9. Move the copied **Item 2** to the corner of the front upright post.

10. Copy the front cross member and move it to the back.

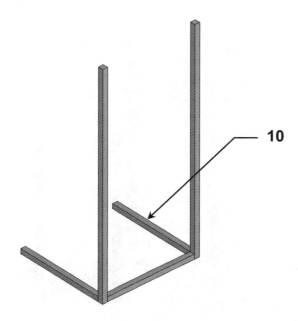

Project 3: Working Platform….continued

11. Select the Copy tool and select the 2 upright posts and the end cross member and copy them to the end of the front cross member as shown below.

12. Copy the front cross member and place it anywhere in the drawing area.

13. Select the copied cross member and change its height in the Properties dialog box to 37.795" [960 mm].

14. Move the copied cross member into the position shown below.

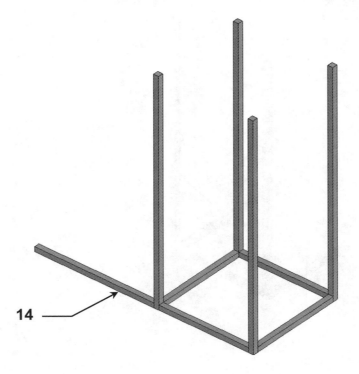

Project 3: Working Platform….continued

15. Copy the longer front cross member and move it to the back.

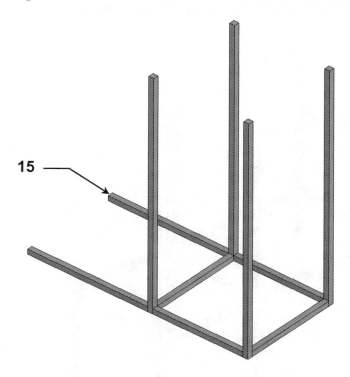

16. Select the Copy tool and select the 2 end upright posts and the end cross member and copy them to the end of the longer front cross member as shown below.

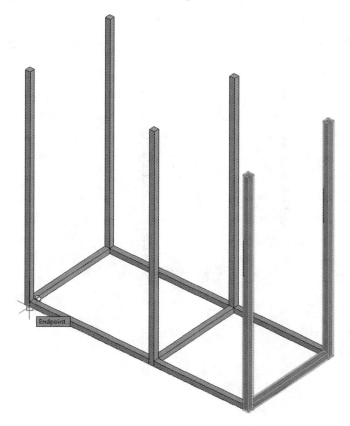

Project 3: Working Platform....continued

17. Select the Move tool then select all the 6 cross members and move them up in the Z axis by 29.921" [760 mm] as shown below. **Note:** Make sure you have Ortho Mode (*F8*) turned on.

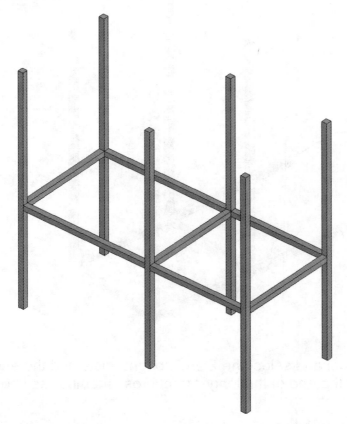

The image below shows the end view of the platform with the dimensions for the position of the cross members.

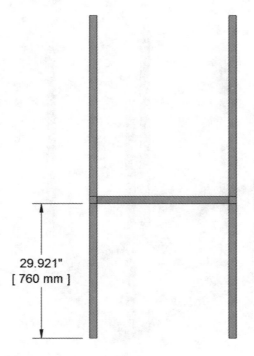

29.921"
[760 mm]

Project 3: Working Platform....continued

The next stage is to create the platform floor.

18. Create a 3D solid Box with the following sizes and place it anywhere in the drawing area.

<u>**Inch users**</u>

Length = 70.866"
Width = 31.496"
Height = 0.787"

<u>**Metric users**</u>

Length = 1800 mm
Width = 800 mm
Height = 20 mm

The next stage is to create the cutouts in the floor to fit around the upright posts. This is achieved by copying the 6 posts and moving them on to the floor, and then by using the Subtract tool.

19. Select the Copy tool and select all 6 upright posts, use the front top corner of one of the posts then move the copied posts to the top corner of the floor as shown below.

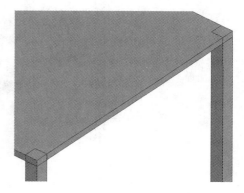

Enlarged view showing the copied upright posts in their positions on the platform floor.

Project 3: Working Platform....continued

20. Select the Subtract tool and subtract the upright posts from the floor.

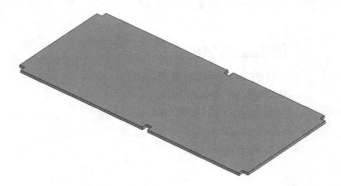

21. Select the Move tool then move the platform floor into position on the main platform cross members as shown below.

22. Resave the drawing file as **Working Platform** but keep the drawing file open.

Project 3: Working Platform....continued

The next stage is to create the cylindrical handrails.

23. Create **Item 4** and place it anywhere in the drawing area.

24. Rotate **Item 4** in the X axis by 90 degrees.

25. Select the Move tool and move the Cylinder to the top of the front left-hand upright posts as shown below. Use the Center Snap on the end of the Cylinder, and the Midpoint Snap on the upright post.

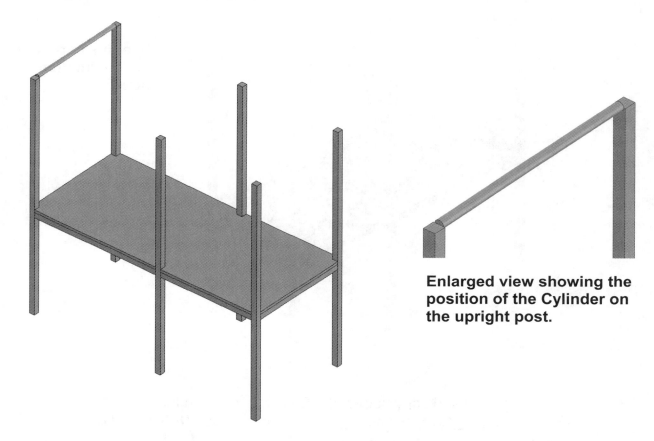

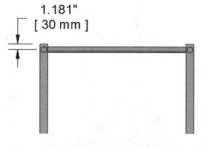

Enlarged view showing the position of the Cylinder on the upright post.

26. Select the Move tool and move the Cylinder down in the Z axis by 1.181" [30 mm].
Note: Make sure you have Ortho Mode (*F8*) turned on.

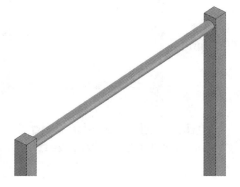

1.181"
[30 mm]

Partial end view showing the position of the Cylinder.

Project 3: Working Platform....continued

27. Select the Copy tool and copy the Cylinder to the right-hand upright posts. Use the Endpoint Snap on the back right-hand edge of the post as the base point, and use the Endpoint Snap on the back right-hand edge of the other upright post as the destination point.

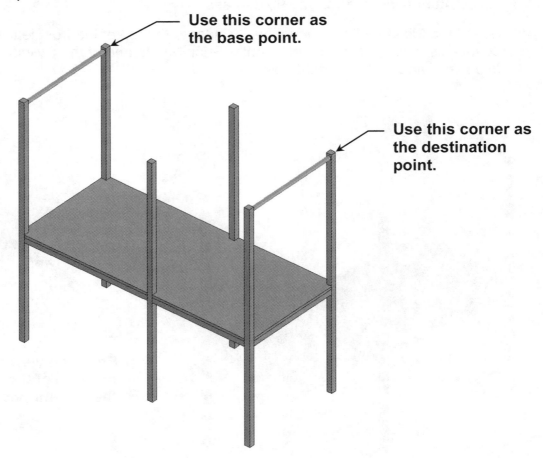

Use this corner as the base point.

Use this corner as the destination point.

28. Select the Copy tool and copy both of the Cylinders down in the Z axis by 19.685" [500 mm]. **Note:** Make sure you have Ortho Mode (*F8*) turned on.

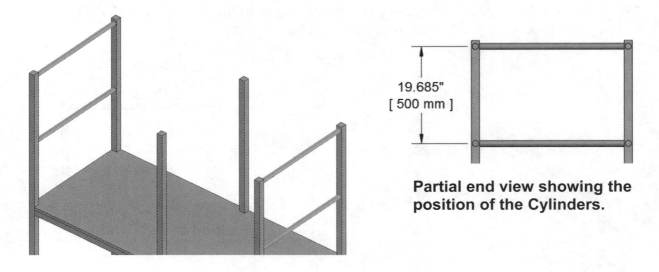

19.685"
[500 mm]

Partial end view showing the position of the Cylinders.

Project 3: Working Platform....continued

29. Copy one of the Cylinders and place it anywhere in the drawing area.

30. Rotate the copied Cylinder by 90 degrees in the Z axis.

31. Move the copied Cylinder into position.

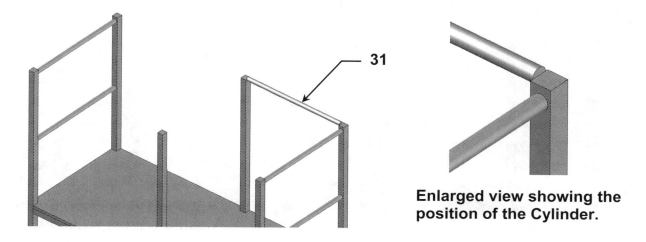

Enlarged view showing the position of the Cylinder.

32. Select the Move tool and move the Cylinder down in the Z axis by 1.181" [30 mm].
    Note: Make sure you have Ortho Mode (**F8**) turned on.

33. Copy the Cylinder and move it down in the Z axis by 19.685" [500 mm]. **Note:** Make sure you have Ortho Mode (**F8**) turned on.

Project 3: Working Platform....continued

34. Create **Item 5** and place it anywhere in the drawing area.

35. Rotate the Cylinder by 90 degrees in the X axis.

36. Rotate the Cylinder by 90 degrees in the Z axis.

37. Move the Cylinder into position.

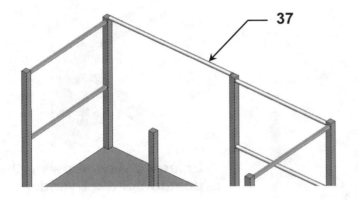

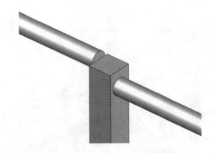

Enlarged view showing the position of the Cylinder.

38. Select the Move tool and move the Cylinder down in the Z axis by 1.181" [30 mm]. **Note:** Make sure you have Ortho Mode (*F8*) turned on.

39. Copy the Cylinder and move it down in the Z axis by 19.685" [500 mm]. **Note:** Make sure you have Ortho Mode (*F8*) turned on.

Project 3: Working Platform....continued

40. Select the Copy tool and copy the 2 longer Cylinders to the front upright posts. Use the Endpoint Snap on the back left-hand edge of the post as the base point, and use the Endpoint Snap on the back left-hand edge of the front upright post as the destination point.

Use this corner as the base point.

Use this corner as the destination point.

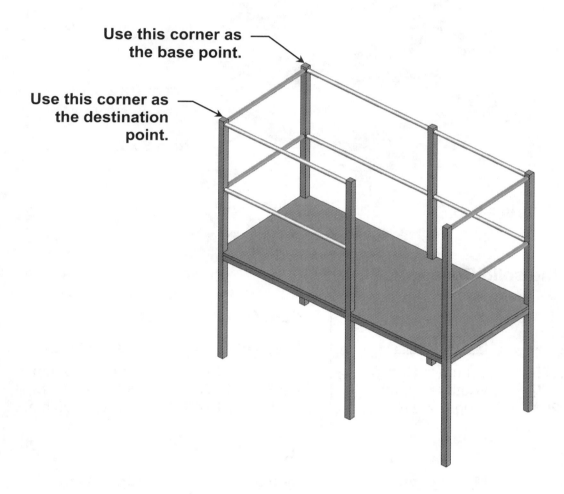

41. Resave the drawing file **Working Platform** but keep the drawing file open.

The main working platform is now complete. The next stage is to create the steps. The sizes for the framework of the steps are shown on the next page.

Project 3: Working Platform....continued

The image below is the main framework for the steps without the step plates, showing the labeled parts and the quantity of each part that is required.

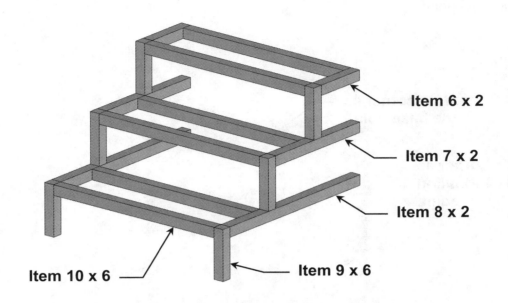

Item 6 x 2

Item 7 x 2

Item 8 x 2

Item 9 x 6

Item 10 x 6

Sizes for the solid Boxes.

Item 6

Inch users

Length = 1.575"
Width = 1.575"
Height = 8.268"

Metric users

Length = 40 mm
Width = 40 mm
Height = 210 mm

Item 7

Inch users

Length = 1.575"
Width = 1.575"
Height = 18.110"

Metric users

Length = 40 mm
Width = 40 mm
Height = 460 mm

Item 8

Inch users

Length = 1.575"
Width = 1.575"
Height = 27.953"

Metric users

Length = 40 mm
Width = 40 mm
Height = 710 mm

Item 9

Inch users

Length = 1.575"
Width = 1.575"
Height = 7.874"

Metric users

Length = 40 mm
Width = 40 mm
Height = 200 mm

Item 10

Inch users

Length = 1.575"
Width = 1.575"
Height = 28.346"

Metric users

Length = 40 mm
Width = 40 mm
Height = 720 mm

Project 3: Working Platform....continued

42. Create 1 **Item 10** and place it anywhere in the drawing area.

43. Rotate **Item 10** in the Y axis by 90 degrees.

44. Create 1 **Item 9** and place it on the end of **Item 10**.

45. Copy the upright leg and move the copy to the other end of **Item 10** as shown below.

46. Create 1 **Item 8** and place it anywhere in the drawing area.

47. Rotate **Item 8** in the X axis by 90 degrees.

48. Move **Item 8** into position as shown below.

49. Copy **Item 8** and move it to the other end as shown below.

50. Copy **Item 10** and move it in the positive Y axis by 8.268" [210 mm]. **Note:** Make sure you have Ortho Mode (*F8*) turned on.

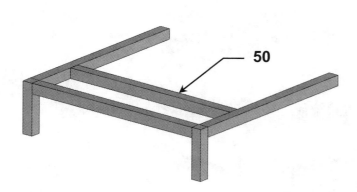

50

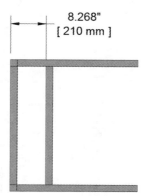

8.268"
[210 mm]

Top view showing the position.

Project 3: Working Platform....continued

51. Copy the 2 upright legs and the 2 front cross members and move them into position as shown below.

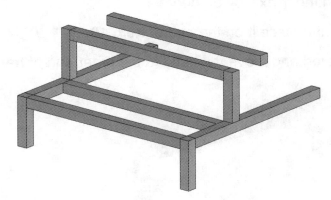

52. Create 1 **Item 7** and place it anywhere in the drawing area.

53. Rotate **Item 7** in the X axis by 90 degrees.

54. Move **Item 7** into position as shown below.

55. Copy **Item 7** and move it to the other end as shown below.

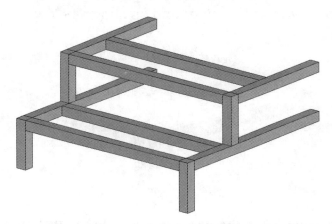

56. Repeat **Step 51** and place them into position as shown below.

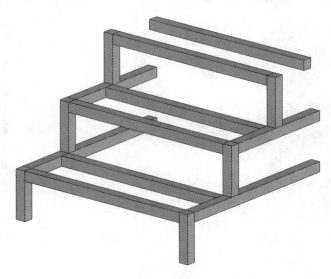

Project 3: Working Platform....continued

57. Create 1 **Item 6** and place it anywhere in the drawing area.

58. Rotate **Item 6** in the X axis by 90 degrees.

59. Move **Item 6** into position as shown below.

60. Copy **Item 6** and move it to the other end as shown below.

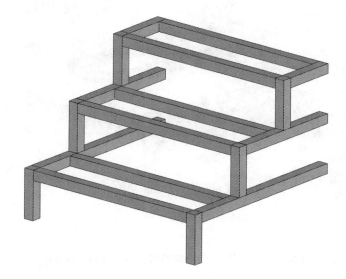

61. Resave the drawing file as **Working Platform** but keep the drawing file open.

The next stage is to create the step plates.

62. Create a 3D solid Box with the following sizes and place it anywhere in the drawing area.

Inch users

Length = 31.496"
Width = 9.843"
Height = 0.787"

Metric users

Length = 800 mm
Width   = 250 mm
Height = 20 mm

Project 3: Working Platform….continued

63. Move the step plate into position as shown below.

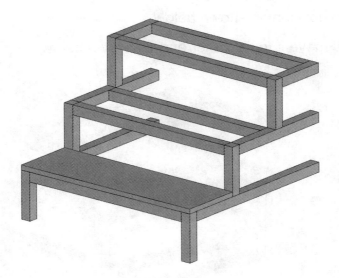

64. Copy the step plate and place 2 copies into their positions as shown below.

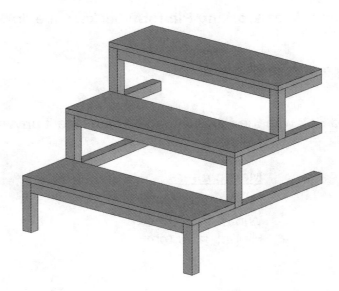

The steps are now complete and the next stage is to move them into position on the main working platform.

Project 3: Working Platform....continued

65. Select the Move tool and select the completed steps using a Window Selection and move the steps to the front right-hand edge of the upright post as shown below.

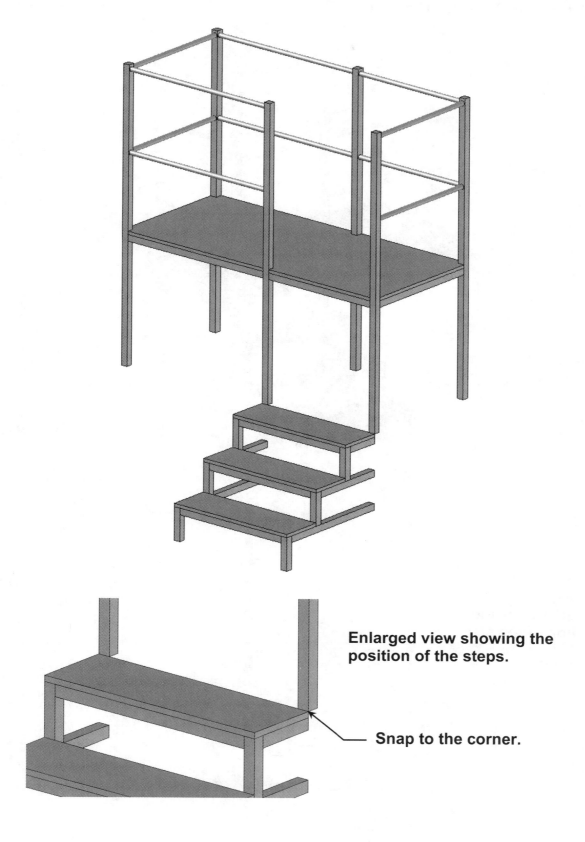

Enlarged view showing the position of the steps.

Snap to the corner.

Project 3: Working Platform....continued

66. Select the Move tool and select the steps using a Window Selection and move the steps up in the Z axis by 24.409" [620 mm] as shown below. **Note:** Make sure you have Ortho Mode (*F8*) turned on.

End view showing the position of the steps.

24.409"
[620 mm]

Project 3: Working Platform....continued

There are various ways you could have created the working platform. You could have used the **Dynamic UCS** to create the Cylinders, or you could have created the cross members using the axis endpoint option of the Box command.

Also try experimenting with different views using the Orbit tool with Perspective view on.

Project 3 is now complete. So resave the drawing file and close it down, then progress on to **Project 4**.

Notes:

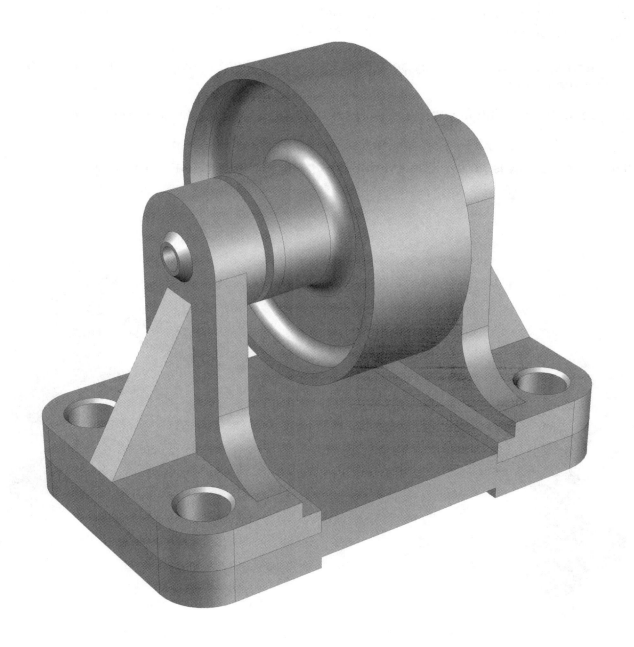

PROJECT 4
Belt Roller Assembly

Project 4: Belt Roller Assembly

In this project you will create 3D solid models for a Belt Roller Assembly. All of the objects are created from 3D solid Boxes and Cylinders.

1. Start a new drawing file by selecting either **acad.dwt** for inch users, or **acadiso.dwt** for metric users.

2. Save the drawing file as **Belt Roller Assembly** but keep the drawing file open.

The image below is an exploded view of the parts required for the assembly.

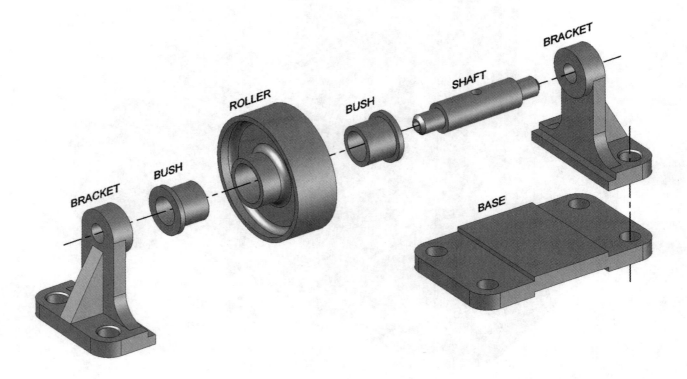

The project will contain dimensioned drawings for users who are familiar with 2D drafting who may wish to create the parts in a different way. The project will also contain detailed instructions on how to create the parts for those users who are not familiar with 2D drafting.

The first stage is to create the base. This will be done using 3 solid Boxes. The reason for this is that one section of the base can be used for the bottom part of the brackets and this approach saves extra work and time.

Project 4: Belt Roller Assembly….continued

Creating the Base.

The dimensioned drawing below shows the base in 3 sections. The first stage is to create **Base 1**, which will then be copied for the bottom of both brackets. The second stage will be to create **Base 3** and then mirror **Base 1** around **Base 3**. The final stage will be to join all 3 sections of the base using the Union tool.

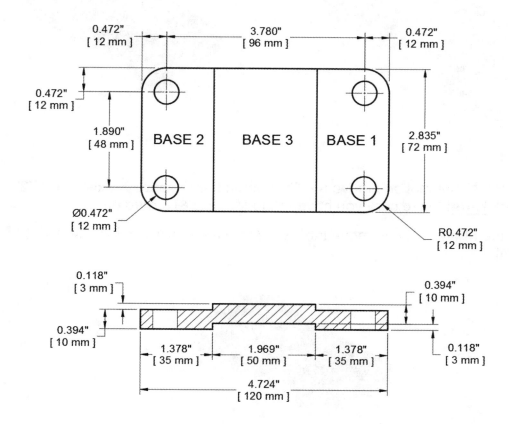

3. Select the **SE Isometric** view.

4. Create a solid Box with the following sizes.

<u>**Inch users**</u>

Length = 1.378"
Width = 2.835"
Height = 0.394"

<u>**Metric users**</u>

Length = 35 mm
Width = 72 mm
Height = 10 mm

Project 4: Belt Roller Assembly....continued

5. Create a solid Cylinder with a radius of 0.236" [6 mm] using the front right-hand bottom corner of the Box as the base point (**P1**), and using the front right-hand top corner of the Box as the height (**P2**).

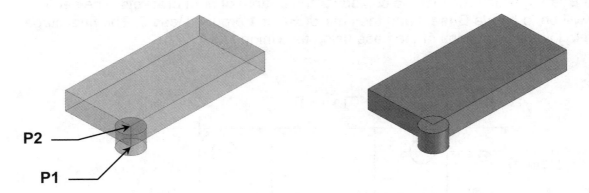

6. Select the **Move** tool and move the Cylinder in the negative X axis by 0.472" [12 mm]. **Note:** Make sure you have Ortho Mode (**F8**) turned on.

7. Select the **Move** tool and move the Cylinder in the positive Y axis by 0.472" [12 mm].

8. Select the **Copy** tool and copy the Cylinder in the positive Y axis by 1.890" [48 mm].

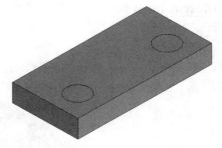

9. Resave the drawing file as **Belt Roller Assembly** but keep the drawing file open.

Project 4: Belt Roller Assembly….continued

10. Select the **Subtract** tool and subtract the 2 Cylinders from the Box.

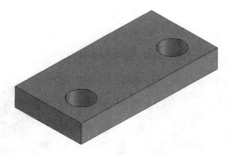

11. Select the **Fillet** tool and fillet the front and back right-hand corners with a fillet radius of 0.472" [12 mm].

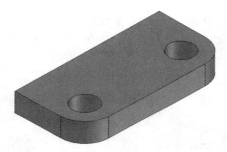

12. Select the Copy tool and place a copy of **Base 1** anywhere in the drawing area. (This will be used later for the bottom of the brackets.)

13. Create a solid Box with the following sizes.

Inch users	Metric users
Length = 1.969"	Length = 50 mm
Width = 2.835"	Width = 72 mm
Height = 0.394"	Height = 10 mm

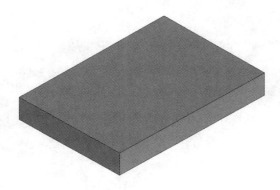

Project 4: Belt Roller Assembly....continued

14. Select the **Move** tool and move **Base 3** to the left-hand end of **Base 1** as shown below. (Use the **Endpoint** Snap.)

15. Select the **3D Mirror** tool and mirror **Base 1** using **P1** as the first point, **P2** as the second point and **P3** as the third point using the **Midpoint** Snap.

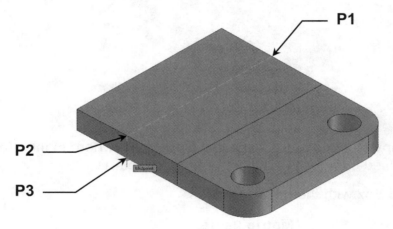

Your models should look like the image below after using the **3D Mirror** tool.

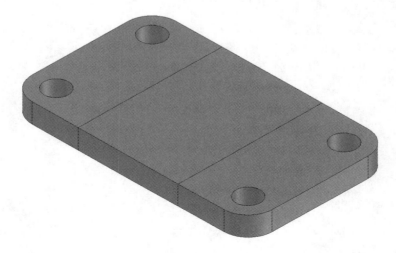

Project 4: Belt Roller Assembly....continued

16. Select the **Move** tool and move **Base 3** up in the positive Z axis by 0.118" [3 mm].
    Note: Make sure you have Ortho Mode (*F8*) turned on.

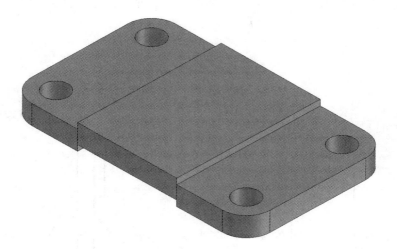

17. Select the **Union** tool and union all 3 sections together.

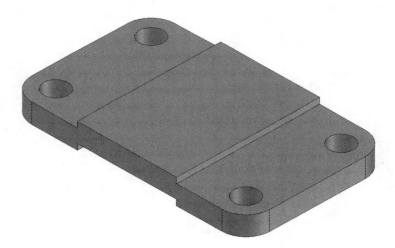

18. Resave the drawing file as **Belt Roller Assembly** but keep the drawing file open.

The base is now complete. The next stage is to create the brackets.

Project 4: Belt Roller Assembly....continued

Creating the Brackets.

The dimensioned drawing below shows the remaining dimensions needed to create the rest of the bracket. The bottom of the bracket was partially created in the first stage.

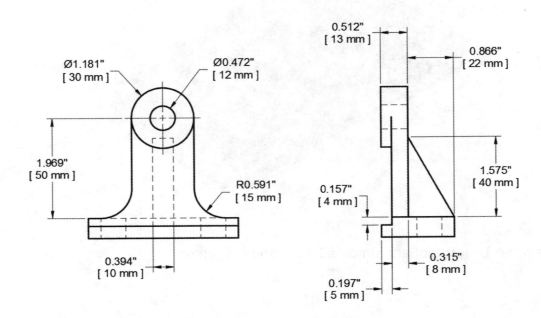

19. Create a solid Box with the following sizes.

Inch users	**Metric users**
Length = 0.197"	Length = 5 mm
Width = 2.835"	Width = 72 mm
Height = 0.157"	Height = 4 mm

Project 4: Belt Roller Assembly....continued

20. Select the **Move** command and move the Box on to **Base 2**, which is the copy of **Base 1** that you created earlier, as shown below.

21. Select the **Subtract** tool and subtract the Box from **Base 2**.

22. Create a solid Box with the following sizes.

Inch users	Metric users
Length = 0.315"	Length = 8 mm
Width = 1.181"	Width = 30 mm
Height = 1.969"	Height = 50 mm

Project 4: Belt Roller Assembly….continued

23. Select the **SW Isometric** view.

24. Select the **Move** tool and move the Box to the midpoint of **Base 2** as shown below. Use the **Midpoint** Snap.

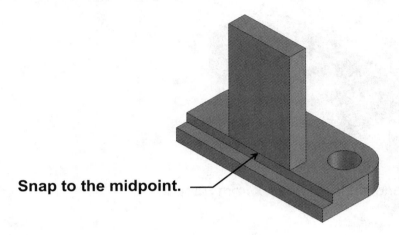

Snap to the midpoint.

25. Select the **SE Isometric** view.

26. Create a solid Cylinder with the following sizes.

Inch users	Metric users
Radius = 0.590"	Radius = 15 mm
Height = 0.512"	Height = 13 mm

27. Select the **3D Rotate** tool and rotate the Cylinder by 90 degrees in the Y axis.

Project 4: Belt Roller Assembly....continued

28. Select the **Move** tool and move the Cylinder to the upright Box using the right-hand center point of the Cylinder as the base point, and the top right-hand midpoint of the upright Box as the destination point.

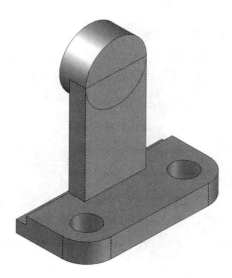

29. Select the **Union** tool and union together the 3 parts of the bracket.

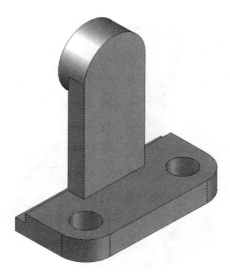

30. Resave the drawing file as **Belt Roller Assembly** but keep the drawing file open.

Project 4: Belt Roller Assembly....continued

31. Select the **Cylinder** tool and use the **Dynamic UCS** to place a Cylinder with a radius of 0.236" [6 mm] on the back upright face using the Center Snap on the center point of the back face radius.

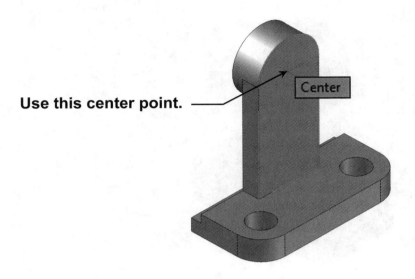

Use this center point. ⸻ Center

32. Create a Cylinder with a length that is longer than is needed as it will be easier to subtract from the main model.

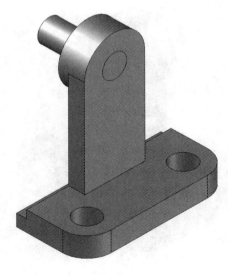

33. Select the **Subtract** tool and subtract the Cylinder from the main model.

Project 4: Belt Roller Assembly....continued

34. Create a solid **Wedge** with the following sizes.

Inch users	Metric users
Length = 0.866"	Length = 22 mm
Width = 0.394"	Width = 10 mm
Height = 1.575"	Height = 40 mm

35. Select the **Move** tool and move the Wedge to the midpoint of the back face.

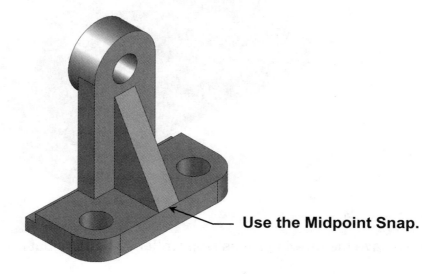

Use the Midpoint Snap.

Project 4: Belt Roller Assembly....continued

36. Select the **Union** tool and union the Wedge to the main model.

37. Select the **Fillet** tool and fillet the 2 bottom corners of the upright, with a fillet radius of 0.590" [15 mm]. (Use the **Orbit** tool to maneuver the model.)

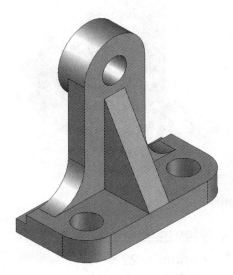

The bracket is now complete. The next stage is to move it on to the base and then create a mirrored copy of the bracket.

38. Select the **Move** tool and place the bracket on to the base as shown below.

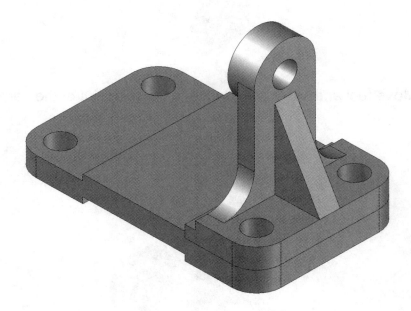

39. Resave the drawing file as **Belt Roller Assembly** but keep the drawing file open.

Project 4: Belt Roller Assembly....continued

40. Select the **SW Isometric** view.

41. Select the **3D Mirror** tool and mirror the bracket using **P1** as the first point, **P2** as the second point and **P3** as the third point using the **Midpoint** Snap.

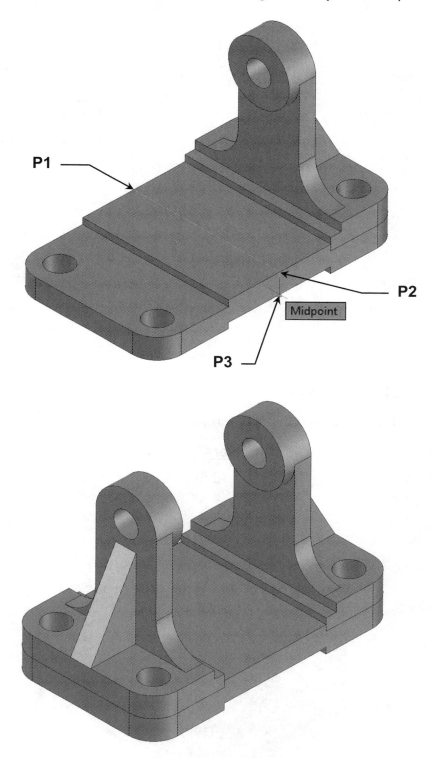

The base and brackets are now complete. The next stage is to create the Roller.

Project 4: Belt Roller Assembly....continued

Creating the Roller.

The dimensioned drawing below is for more experienced users who wish to create the Roller in their own way.

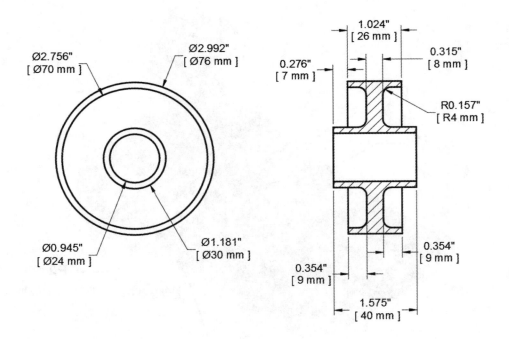

42. Select the **SE Isometric** view.

43. Create a solid Cylinder with the following sizes.

Inch users	**Metric users**
Radius = 1.496"	Radius = 38 mm
Height = 1.024"	Height = 26 mm

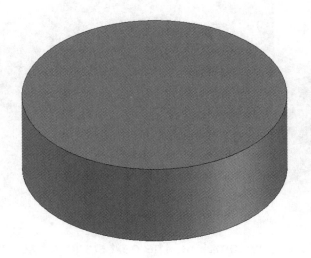

Project 4: Belt Roller Assembly....continued

44. Select the Cylinder and change its Transparency to 70 in the **Properties** dialog box.

45. Create another Cylinder with the following sizes using the bottom center point of the first Cylinder as the base point.

   Inch users

   Radius = 1.378"
   Height = 0.354"

   Metric users

   Radius = 35 mm
   Height = 9 mm

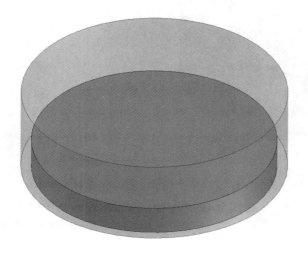

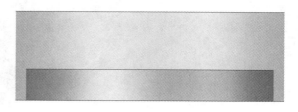

Front view showing the position of the new Cylinder.

46. Select the **Copy** tool and copy the smaller Cylinder to the top of the larger Cylinder as shown below. (Use the **Center** Snap on the top face of the smaller Cylinder for the base point, and the **Center** Snap on the top face of the larger Cylinder for the destination point.) **Note:** You may need to use a Window Selection to select the smaller Cylinder.

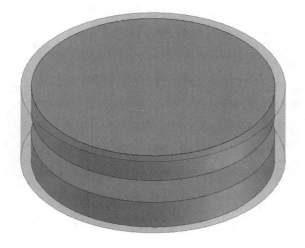

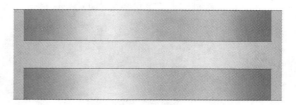

Front view showing the position of the new Cylinder.

Project 4: Belt Roller Assembly....continued

47. Select the **Subtract** tool and subtract the 2 smaller Cylinders from the main roller.

48. Change the transparency of the roller back to 0.

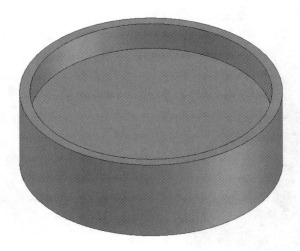

49. Create another solid Cylinder with the following sizes and place the center base point on the center of the top hole in the larger Cylinder. **Note:** Be careful when placing the base point of the Cylinder. You will see 4 center locations. The one you require is the second one down from the top.

Inch users	**Metric users**
Radius = 0.590"	Radius = 15 mm
Height = 1.575"	Height = 40 mm

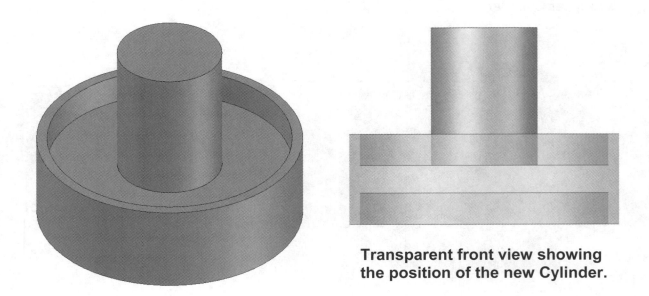

Transparent front view showing the position of the new Cylinder.

Project 4: Belt Roller Assembly....continued

50. Select the **Move** tool and move the new Cylinder down in the negative Z axis by 0.945" [24 mm]. **Note:** Make sure you have Ortho Mode (*F8*) turned on.

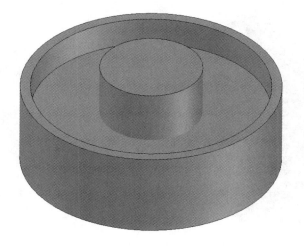

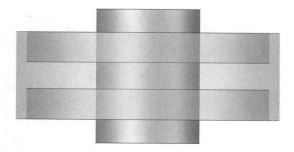

Transparent front view showing the position of the new Cylinder.

51. Select the **Union** tool and union the small Cylinder to the main roller.

52. Create another Cylinder with the following sizes using the top center point on the main roller as the base point. **Note:** After entering the radius, move the mouse cursor down in the negative Z axis before entering the height.

Inch users

Radius = 0.472"
Height = 1.575"

Metric users

Radius = 12 mm
Height = 40 mm

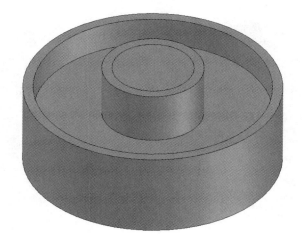

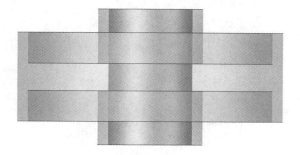

Transparent front view showing the Cylinder.

Project 4: Belt Roller Assembly....continued

53. Select the **Subtract** tool and subtract the Cylinder from the main roller.

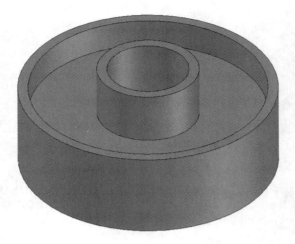

54. Select the **Fillet** tool and fillet the 4 internal edges with a fillet radius of 0.157"
    [4 mm]. Use the **Orbit** tool to place the 2 fillets on the underside of the roller.

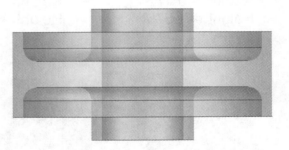

**Transparent front view showing
the 4 fillets.**

Project 4: Belt Roller Assembly....continued

55. Select the **3D Rotate** tool and rotate the roller in the Y axis by 90 degrees.

56. Resave the drawing file as **Belt Roller Assembly** but keep the drawing file open.

The Roller is now complete. The next stage is to create the 2 Bushes.

Creating the Bushes.

The dimensioned drawing below is for more experienced users who wish to create the Bushes in their own way.

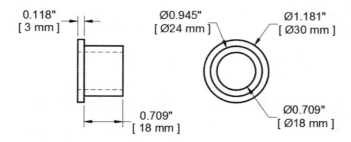

57. Create a solid Cylinder with the following sizes.

Inch users	Metric users
Radius = 0.590"	Radius = 15 mm
Height = 0.118"	Height = 3 mm

Project 4: Belt Roller Assembly….continued

58. Create another solid Cylinder with the following sizes and place the base point on the top center point of the previous Cylinder.

Inch users

Radius = 0.472"
Height = 0.709"

Metric users

Radius = 12 mm
Height = 18 mm

59. Select the **Union** tool and union both Cylinders together.

60. Change the transparency of the bush to 70.

61. Create another Cylinder with a radius of 0.354" [9 mm] using the bottom center point of the bush as the base point, and the top center point of the bush for the height.

Transparent front view showing the new Cylinder.

62. Select the **Subtract** tool and subtract the Cylinder from the main bush.

Project 4: Belt Roller Assembly....continued

63. Select the **3D Rotate** tool and rotate the bush 90 degrees around the Y axis.

64. Select the **Copy** tool and copy the bush. Place it anywhere in the drawing area.

65. Select the **3D Rotate** tool and rotate the copied bush 180 degrees around the Z axis.

66. Use the **Move** tool to move the bushes and to place them into position on the roller. **Note:** Make sure you use the correct **Center** Snap on the bush and on the roller.

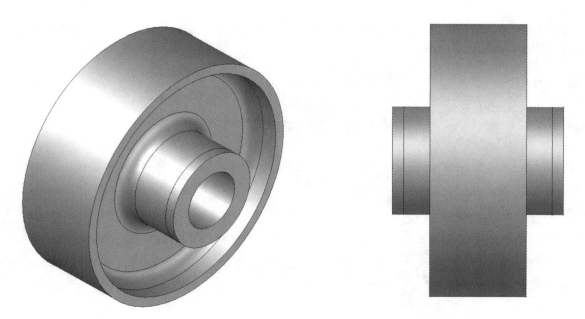

The Bushes are now complete and placed into position on the Roller. The next stage is to create the Shaft.

Project 4: Belt Roller Assembly....continued

Creating the Shaft.

The dimensioned drawing below is for more experienced users who wish to create the Shaft in their own way.

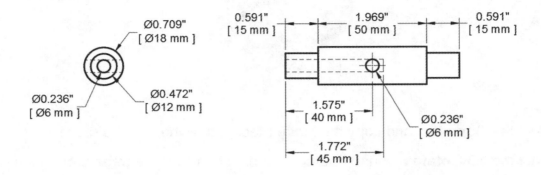

67. Create a solid Cylinder with the following sizes.

Inch users

Radius = 0.236"
Height = 0.591"

Metric users

Radius = 6 mm
Height = 15 mm

68. Create another Cylinder with the following sizes and place it on top of the first one using the **Center** Snap.

Inch users

Radius = 0.354"
Height = 1.969"

Metric users

Radius = 9 mm
Height = 50 mm

Project 4: Belt Roller Assembly....continued

69. Create another Cylinder with the following sizes and place it on top of the second one using the **Center** Snap.

Inch users

Radius = 0.236"
Height = 0.591"

Metric users

Radius = 6 mm
Height = 15 mm

70. Select the **Union** tool and union all 3 Cylinders together.

71. Select the Shaft and change the Transparency to 70.

72. Create another Cylinder with the following sizes using the top center point on the Shaft as the base point. **Note:** After entering the radius, move the mouse cursor down in the negative Z axis before entering the height.

Inch users

Radius = 0.118"
Height = 1.772"

Metric users

Radius = 3 mm
Height = 45 mm

Project 4: Belt Roller Assembly....continued

73. Select the **Subtract** tool and subtract the Cylinder from the main Shaft.

74. Select the **3D Rotate** tool and rotate the Shaft by 90 degrees in the Y axis..

75. Create another Cylinder with the following sizes and place it anywhere in the drawing area.

Inch users

Radius = 0.118"
Height = 1.181"

Metric users

Radius = 3 mm
Height = 30 mm

76. Change the transparency of the Shaft back to 0.

77. Select the **Move** tool and move the Cylinder using the top center point as the base point, and the center point at the right-hand of the Shaft as shown below.

Project 4: Belt Roller Assembly....continued

78. Select the **Move** tool and move the Cylinder up in the positive Z axis by 0.591"
    [15 mm] as shown below. **Note:** Make sure you have Ortho Mode (*F8*) turned on.

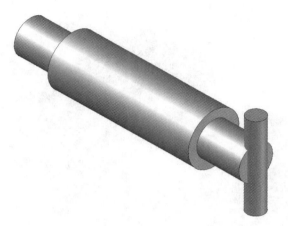

79. Select the **Move** tool and move the Cylinder up in the negative X axis by 1.575"
    [40 mm] as shown below.

80. Select the **Subtract** tool and subtract the Cylinder from the main Shaft.

81. Resave the drawing file as **Belt Roller Assembly** but keep the drawing file open.

The Shaft is now complete. The next stage is to move the Shaft on to the Roller and
Bushes.

Project 4: Belt Roller Assembly….continued

82. Select the **Move** tool and move the Shaft into position on the right-hand bush using the **Center** Snap, as shown below.

83. Select the **Move** tool and move the Shaft in the positive X axis by 0.079" [2 mm] as shown below. **Note:** Make sure you have Ortho Mode (*F8*) turned on.

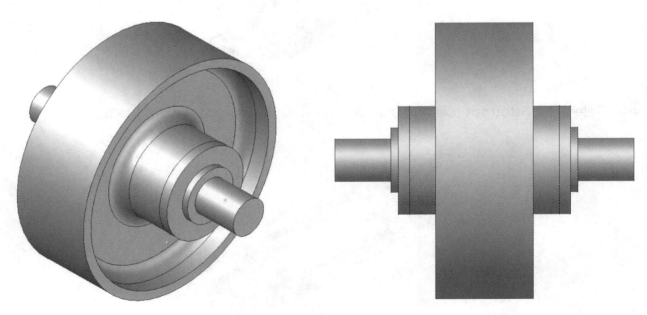

Front view showing the final position of the Shaft.

The next stage is to move the Brackets and Base on to the Shaft, Bushes and Roller.

Project 4: Belt Roller Assembly....continued

84. Select the **Move** tool then select both Brackets and the Base. Use the center point for the base point (**P1**), and move the Brackets and Base on to the Shaft using the center point for the destination point (**P2**).

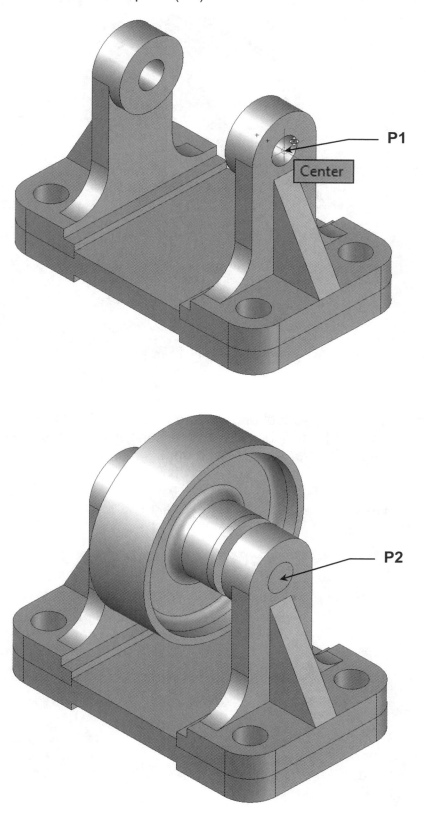

Project 4: Belt Roller Assembly....continued

85. Select the **Move** tool then select both Brackets and the Base and move them in the negative X axis by 0.079" [2 mm] as shown below. **Note:** Make sure you have Ortho Mode (**F8**) turned on.

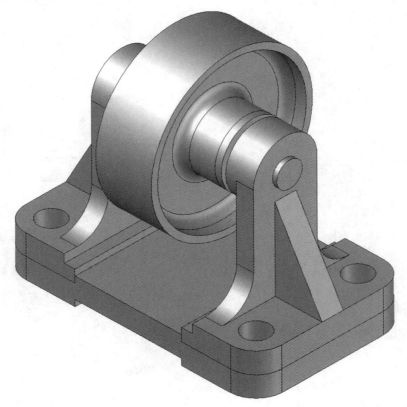

86. Select the **Chamfer Edge** tool and place a 0.078" x 0.078" [2 mm x 2 mm] chamfer on each end of the Shaft. Chamfer the right-hand end first then switch to **SW Isometric** view to chamfer the left-hand end.

Front view showing both chamfers.

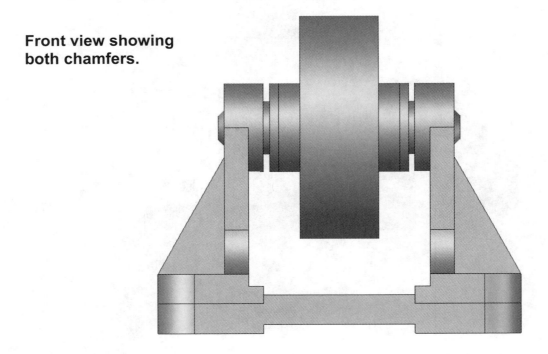

Project 4: Belt Roller Assembly....continued

87. Select the **Chamfer Edge** tool and place a 0.040" x 0.040" [1 mm x 1 mm] chamfer on the top edges of the holes in the Brackets.

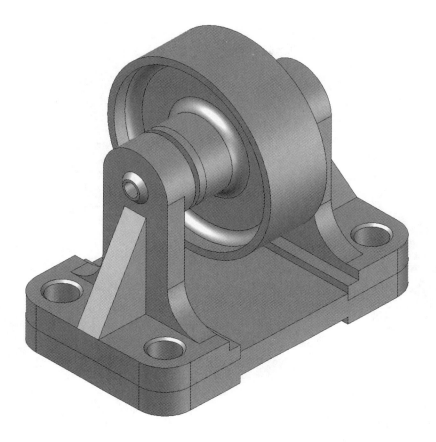

88. Resave the drawing file as **Belt Roller Assembly**.

Project 4 is now complete. Experiment with changing the colors of the components.

Notes:

Appendix A: 3D Printing

AutoCAD has 2 methods that you can use to send your 3D models to a 3D printer. You can choose to save your 3D model as an **STL** file, which you can then send to a 3D printing service. You can also send your 3D model directly to a 3D printer that is connected to your computer using AutoCAD's "**Print Studio**".

Note: Print Studio is not loaded by default and needs to be downloaded from Autodesk's website. Print Studio is **only** available for **AutoCAD versions 2017 for 64-bit systems or later**. Refer to the instructions on the following pages on how to download Print Studio.

How to save your 3D model as an STL file.

1. Select the **3DPRINTSERVICE** command.

   Application Menu / Publish / Send to 3D Print Service
   or
   Keyboard = 3dprintservice <enter>

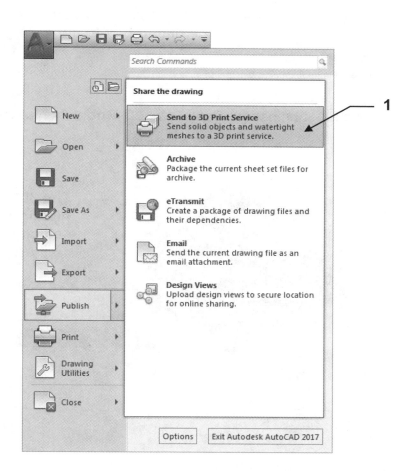

Continued on the next page...

Appendix A: 3D Printing....continued

2. The **3D Printing - Prepare Model for Printing** dialog box will open.

   You can choose to **Learn about preparing a 3D model for printing** by selecting the link in the dialog box. Or you can choose **Continue**. For this exercise we will choose Continue.

3. Select **Continue**.

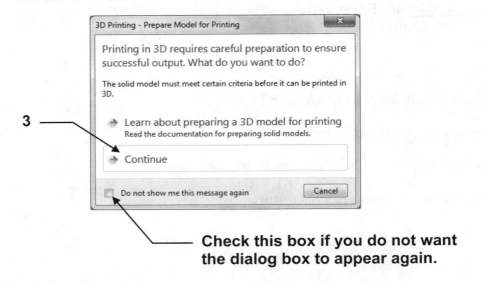

Check this box if you do not want the dialog box to appear again.

4. Select solids or watertight meshes: *Select the 3D Model* then press *<enter>*

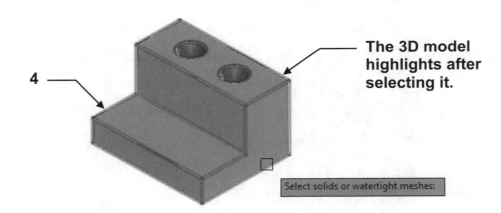

The 3D model highlights after selecting it.

Continued on the next page...

Appendix A: 3D Printing....continued

5. The **3D Print Options** dialog box will open.

You can change the **Scale**, **Length**, **Width** or **Height** by changing any of the **Output dimensions**. It doesn't matter which output dimension you change, as all the other dimensions will alter proportionately.

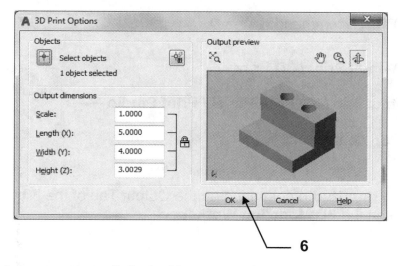

6

6. Select **OK** when you are satisfied with your settings.

7. The **Create STL File** dialog box will open.

8. Type a name for your **STL** file then choose a location to save it in.

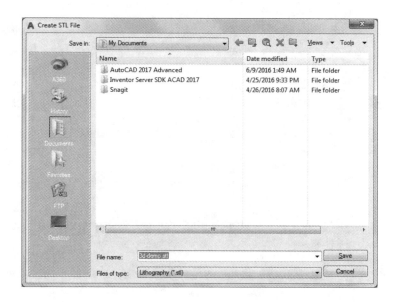

Your 3D model is now saved as an STL file, which can be sent to a 3D printing service, or you can use the STL file for your own 3D printer.

Continued on the next page...

Appendix A: 3D Printing....continued

How to print your 3D model on a 3D printer.

Note: Print Studio is only available for **AutoCAD versions 2017 for 64-bit systems or later**.

1. Select the **3DPRINT** command.

   Application Menu / Print / 3D Print
   or
   Ribbon = Output Tab / 3D Print Panel / Print Studio ⟶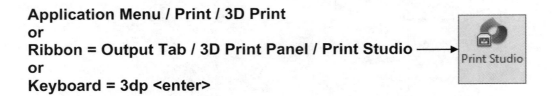
   or
   Keyboard = 3dp <enter>

Note: To use the **Print Studio** command on the Output Tab of the Ribbon, you must make sure you are in the **3D Modeling Workspace**.

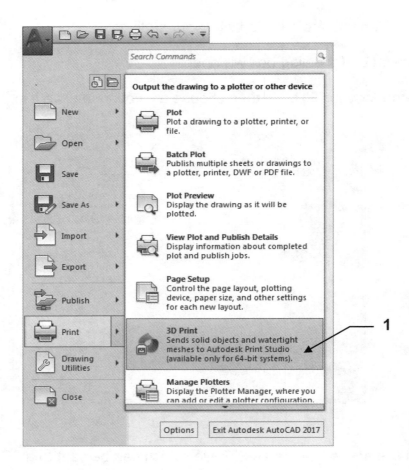

Continued on the next page...

Appendix A: 3D Printing....continued

2. The **3D Printing - Print Studio Not Installed** dialog box will open.

3. Select "**Install Print Studio**". (**Note:** You must be connected to the internet. If you select "**Do not install at this time**" the dialog box will close and the 3D Print command will end.)

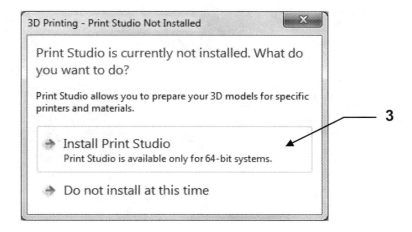

3

4. Follow the instructions for downloading and installing **Print Studio**.

5. The **3D Printing - Prepare Model for Printing** dialog box will open.

6. Select **Continue**.

7. Select solids or watertight meshes: *Select the 3D Model* then press *<enter>*

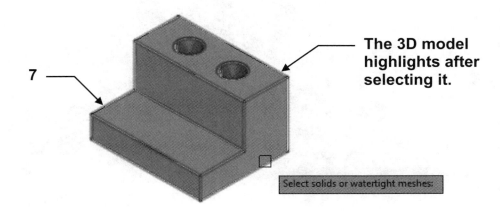

7

The 3D model highlights after selecting it.

Select solids or watertight meshes:

Continued on the next page...

Appendix A: 3D Printing....continued

8. The **3D Print Options** dialog box will open.

You can change the **Scale**, **Length**, **Width** or **Height** by changing any of the **Output dimensions**. It doesn't matter which output dimension you change, as all the other dimensions will alter proportionately.

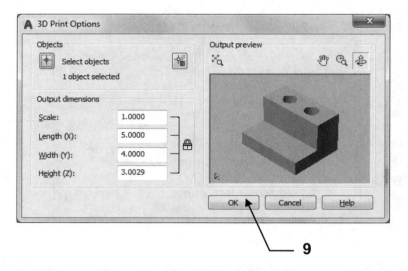

9

9. Select **OK** when you are satisfied with your settings.

10. The **Print Studio** dialog box will open.

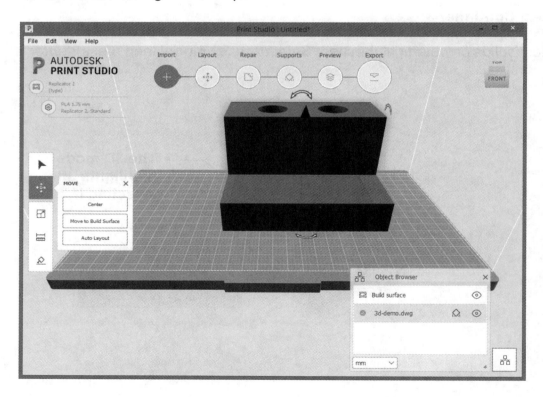

Continued on the next page...

Appendix A: 3D Printing….continued

11. There are various settings you can change in the **Print Studio** dialog box. Some basic ones are explained below.

   a. If your system is connected to a 3D printer the **Print** button will be active. If your system is not connected to a 3D printer the **Export** button will be active, and you can save the 3D model as a 3D printer file for sending to a print service or for later use.

   b. You can preview your model by selecting the **Preview** button, then entering the number of slices in the dialog box.

   c. You can choose a different 3D printer or add a 3D printer by selecting the button. (**Note:** If you are adding a 3D printer it must be connected to your system.)

12. When you are satisfied with all the settings select either **Print** or **Export**.

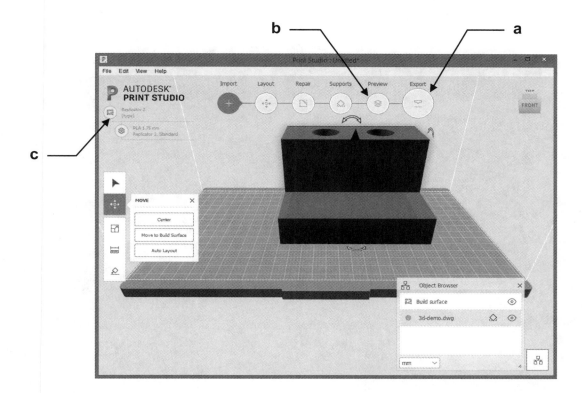

Notes:

Appendix B: Add a Printer / Plotter

The following are step-by-step instructions on how to configure AutoCAD for your printer or plotter. These instructions assume you are a single system user. If you are networked or need more detailed information please refer to your AutoCAD Help Index.

Note: You can configure AutoCAD for multiple printers. Configuring a printer makes it possible for AutoCAD to display the printing parameters for that printer.

1. Type in *plottermanager* and then press **<enter>**.

2. Select **Add-a-Plotter Wizard**.

3. Select the **Next** button.

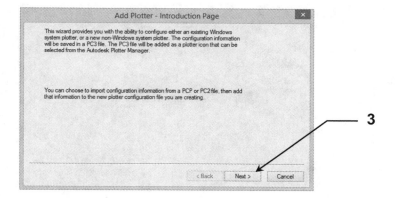

4. Select **My Computer** and then select **Next**.

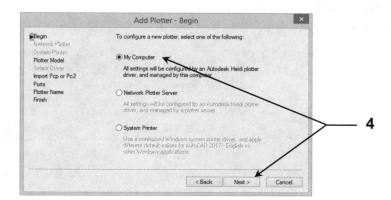

Continued on the next page...

Appendix B: Add a Printer / Plotter....continued

5. Select the **Manufacturer** and the specific **Model** desired then **Next**.

    (If you have a disk with the specific driver information put the disk in the disk drive and select "Have disk" button then follow instructions.)

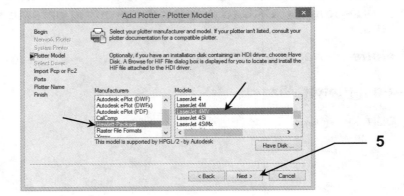

6. Select the **Next** button.

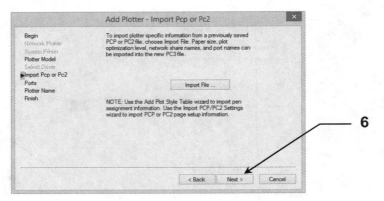

7. Select:
    a. **Plot to a port**.
    b. Then select **Next**.

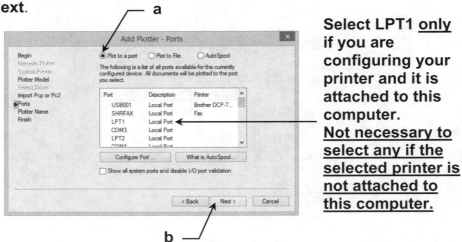

Select LPT1 <u>only</u> if you are configuring your printer and it is attached to this computer. <u>Not necessary to select any if the selected printer is not attached to this computer.</u>

Continued on the next page...

8. The Printer name that you previously selected should appear. Then select the **Next** button.

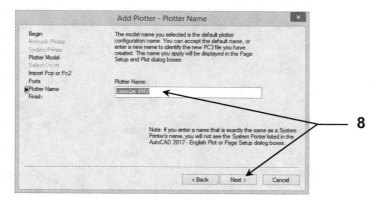

9. Select the **Edit Plotter Configuration** button.

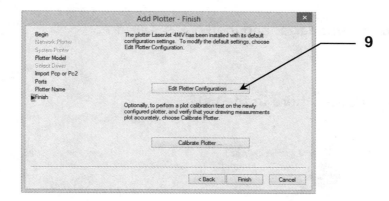

10. Select:
    a. Device and Document Settings Tab.
    b. Media: Source and Size.
    c. Size: (Select the appropriate size for your Printer / Plotter.)
    d. Select the **OK** button.

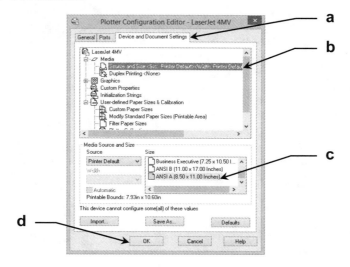

Continued on the next page...

Appendix B: Add a Printer / Plotter....continued

11. Select the **Finish** button.

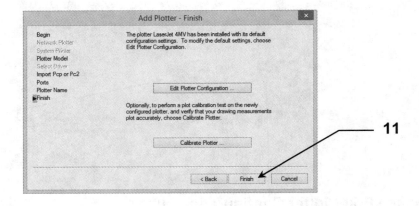

11

12. Type in *plottermanager* and then press *<enter>*.

Is the Printer / Plotter there in the list?

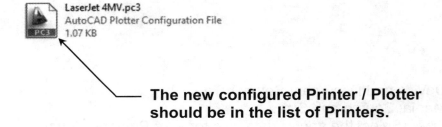

The new configured Printer / Plotter should be in the list of Printers.

INDEX

2D

2D Drafting & Annotation Workspace, Intro-14
2D Model Space, Intro-5
2D Path, 7-19
2D Wireframe, Intro-9, 1-9

3D

3D Align Tool, 6-8
3D Mirror Tool, 6-6
3D Modeling Workspace, Intro-14, 1-3
3D Printing, Appendix-A1
3D Rotate Tool, 6-2
3D Tool, selecting a, 1-2
3D Tools Tab, 1-2

A

About, Author, Intro-1
About, Workbook, Intro-1
Add a Printer / Plotter, Appendix-B1
Align, 3D, 6-8
Align the Origin, 4-7
Axis Endpoint, 1-12

B

Boolean Panel, 5-7
Box, creating a solid, 1-6

C

Chamfer, modify existing, 3-8

Chamfer, remove existing, 3-10
Chamfer Edge Tool, 3-3
Changing the Drawing View, Intro-7
Changing the Visual Style, Intro-9
Changing the Workspace Colors, Intro-4
Circumscribed, pyramid, 1-16
Closed Shapes, create, 7-5
Color, changing, Project 1-13
Color Books, Project 1-14
Color Scheme, Intro-4
Command Line, Intro-14, 1-5
Conceptual Style, Intro-9
Cone, creating a solid, 1-13
Continuous Orbit, 5-2
Copy Tool, 5-9
Cross Sections, lofting, 7-15
Cursor, mouse, Intro-14
Customization Button, 4-10
Cylinder, creating a solid, 1-10

D

Delete key, 3-10
Diameter, entering, 1-15
Drafting and Annotation Workspace, 1-2
Drafting Settings, dialog box, 4-3
Draw Panel, Intro-14
Draw pull-down Menu, 1-4
Drawing File, starting a new, Intro-3
Drawing View, changing, Intro-7
Dynamic Input, Intro-14, 1-5
Dynamic UCS, 4-10

E

Edit Polyline Tool, 7-4
Equal Chamfer, 3-5
Extrude Tool, 7-2

F

Fillet, modify existing, 3-15
Fillet, remove existing, 3-17
Fillet Edge Tool, 3-11
Free Orbit, 5-2
Function keys, keyboard, Intro-12

G

Grid Display, Intro-10
Grid Lines, Intro-10

H

Helix Tool, 8-6
Home Tab, Intro-14
Horizontal Movement, cursor, Intro-11

I

Inscribed, pyramid, 1-17
Intersect Tool, 5-16

K

Keyboard entry, 1-5
Keyboard Function keys, Intro-12

L

Lighting, highlight intensity, Project 1-17
Loft Tool, 7-13

M

Menus, Workspace, Intro-14
Modify an existing Chamfer, 3-8
Modify an existing Fillet, 3-15
Modify drop-down Panel, 7-4
Modify Solids, properties palette, 2-4
Modify Solids, using grips, 2-6
Move the Origin, 4-6
Move Tool, 5-4
Moving the UCS, 3 point command, 4-4

N

Navigate Panel, 5-2
Navigation Bar, 5-3

O

Object Snap, drafting settings, 4-3
Object Snap, Intro-11
Object Snap icon, 4-3
Object Snap Menu, Intro-13, 5-24
Orbit Tool, 5-2
Origin, move, 4-6
Origin icon, Intro-14
Orthographic (Ortho) Mode, Intro-11
Output Tab, Appendix-A4

P

Pan, wheel mouse, Intro-13
Panel, Intro-14
Path, 2D, 7-19
Polyline, edit, 7-4
Print Studio, Appendix-A4
Printer / Plotter, add, Appendix-B1
Properties Palette, opening, 2-2
Properties Palette, transparency, 5-26
Pyramid, creating a solid, 1-16

Q

Quadrant Snap, 7-20
Quick Access toolbar, 1-4

R

Realistic style, Intro-9
Remove an Existing Chamfer, 3-10
Remove an Existing Fillet, 3-17
Revolve Tool, 7-9
Ribbon, Workspace, Intro-14
Right click Menu, 2-3
Rotating the UCS, 4-8

S

Sample files, Intro-1
Select Color, dialog box, Project 1-13
Select Template, Intro-3
Selecting a Basic 3D Tool, 1-2
Shell Tool, 8-2
Shift key, 5-24
Show Menu Bar, 1-4
Sketchy, Visual Style, Project 1-17
Solid History System Variable, 3-2
South East Isometric View, Intro-7
Sphere, creating a solid, 1-15
Spiral, Helix Tool, 8-6
Start AutoCAD, Intro-2
Starting a new drawing file, Intro-2
Status Bar, Intro-10
Status Bar Buttons, Intro-11
Status Bar Menu, 4-10
Status Bar Tools, Intro-14
STL file, 3D printing, Appendix-A1
Style, conceptual, Intro-9
Subtract Tool, 5-12
Sweep Tool, 7-17

T

Tab, Workspace, Intro-14

Temporary Origin, UCS, 4-2

Temporary Origin, UCS, 4-2
Text labels, status bar, Intro-12
Three-dimensional. *See* 3D
Tool Buttons, status bar, Intro-12
Toolbars, Workspace, Intro-14
Torus, creating a solid, 1-22
Transparency, 5-26
Two-dimensional. *See* 2D

U

UCS, Dynamic, 4-10
UCS, moving, 4-4
UCS, rotating, 4-8
Unequal Chamfer, 3-7
Union Tool, 5-6
User Coordinate System (UCS), 4-2

V

Vertical movement, cursor, Intro-11
View Tab, 5-2
Visual styles, Intro-9, Project 1-16
Visual Styles Manager, Project 1-15

W

WCS, 4-2
Wedge, creating a solid, 1-18
Wheel mouse, using, Intro-13
Wireframe, 2D, Intro-9, 1-9
Workspace, Intro-3
Workspace Colors, changing, Intro-4
Workspace descriptions, Intro-14
Workspace Ribbon, Intro-14
World Coordinate System (WCS), 4-2

Z

Zoom, wheel mouse, Intro-13
Zoom Extents, Intro-13

Notes:

Notes:

Notes:

Notes:

Final Notes about AutoCAD®

Whether a student, teacher, professional, or home user, you can make efficiency a daily part of the task at hand with AutoCAD® software. Meticulously refined with the drafter in mind, AutoCAD facilitates efficient day-to-day drafting with features that increase speed and accuracy while saving time

For information on available trial versions and to purchase AutoCAD software, please visit Autodesk's web page at http://www.autodesk.com.

If you are attending a school, check with your instructor regarding potential educational discounts.

Sample files accompanying this book

To get the supplied drawings that go with this book, download the file:

3d-modeling.zip

from our website:

http://new.industrialpress.com/ext/downloads/acad/3d-modeling.zip

Enter the address into your web browser and the download will start automatically. Once the file has been downloaded, you can unzip it to extract the sample files.

For More Information

For a list of other titles in this bestselling series of AutoCAD Workbooks published by Industrial Press, see page iii at the front of this book.

For further information about these and other Industrial Press products, please visit us at **http://industrialpress.com** and **http://ebooks.industrialpress.com**.